MASTERING MECHANICS I;
USING MATLAB5

Douglas W. Hull

Prentice Hall
Upper Saddle River, New Jersey 07458

Library of Congress Cataloging-in-Publication Data

Hull, Douglas W.
    Mastering mechanics I using MATLAB 5: a guide to statics and
strengths of materials/ Douglas W. Hull.
        Includes index.
        ISBN 0-13-864034-3
        1. MATLAB. 2. Engineering mathematics—Data processing.
    3. Mechanics, Applied—Data processing. I. Title. II. Series.
TA345.H85 1999   98–41443
620.1'01'5118—dc21   CIP

Editorial/Production Supervision: Ann Marie Kalajian
Managing Editor: Eileen Clark
Publisher: Tom Robbins
Acquisitions Editor: Alice Dworkin
Manufacturing Manager: Pat Brown
Assistant Vice President of Manufacturing and Production: David Riccardi
Cover Design Director: Jayne Conte
Cover Design: Karen Salzbach
Editorial Assistant: Dan DePasquale

The author and publisher of this book have used their best efforts in preparing this book. These efforts include the development, research, and testing of theories and programs to determine their effectiveness. The author and publisher shall not be liable in any event for incidental or consequential damages in connection with, or arising out of the furnishing, performance, or use of these formulas.

Printed in the United States of America
10 9 8 7 6 5 4 3 2 1

ISBN   0-13-864034-3

Prentice-Hall International (UK) Limited, *London*
Prentice-Hall of Australia Pty. Limited, *Sydney*
Prentice-Hall Canada Inc., *Toronto*
Prentice-Hall Hispanoamericana, S.A., *Mexico*
Prentice-Hall of India Private Limited, *New Delhi*
Prentice-Hall of Japan, Inc., *Tokyo*
Simon & Schuster Asia Pte. Ltd., *Singapore*
Editora Prentice-Hall do Brasil, Ltda., *Rio de Janeiro*

Trademark Information
*MATLAB is a registered trademark
of the MathWorks, Inc.*
The MATHWORKS, Inc.
24 Prime Park Way
Natick, Mass 01760-1500
Phone: (508) 647-7000
Fax: (508) 647-7001
Email: info@mathworks.com
http://www.mathworks.com

*"To those who
carry a candle in the darkness,
who defend the rights of the individual,
who live the dream."*

Contents

Preface

The purpose of this book is twofold. It presents a usable toolbox for the most common statics and strength of materials problems; it also shows by example how to create function files to solve generic problems in any area. These function files expand the usability if MATLAB® into new areas of study.

This book is designed for those who are currently enrolled in a mechanics course, or have already completed one. The theory presented here is sufficient to remind the reader of the formulas and their assumptions, but the book is not intended to be a primary teaching tool. It is best used as a review or supplemental text.

The reader needs only a passing familiarity with MATLAB itself. It is quite possible to use this text with little familiarity of the program. MATLAB skills will be refined by using this book .

All the function files and sample problems have been made available free of charge via file transfer protocol (ftp) from the MathWorks' World Wide Web site. The files may be found at:

ftp://ftp.mathworks.com/pub/books/hull

The book can be read on two levels. The reader who is interested primarily in obtaining answers to problems, without exploring the inner workings of functions, should look at the examples following each of the code templates and should study the example problems. If the reader places more emphasis on the theory and technical aspects of the calculations, the theory section will be of greater interest. The code for each of the functions is included for those who are interested in coding technique.

The sequence of chapters can be changed, with the exception of the first chapter on statics which should at least be skimmed before moving on to the rest of the text. Chapters are cross-referenced for ease in locating techniques from other chapters.

Acknowledgments: The author would like to thank the many people who offered advice and moral support along the way: Dr. Carl Vilmann, Dr. KVC Rao, Dr. Chris Passerello, Ellen Adams, Michael Agostini, James Chye, Myles Dexter, Matthew Monte, Jennifer Pearson, C.O. Rudstrom, Scott Sherrill, Neal Stangis, Christine Williams, the CAEL partners and anyone else who gave a piece of technical advice in passing as this text was being completed.

On the publication side, the author would like to thank the publisher Tom Robbins, his former assistant Nancy Garcia, the other people who have helped: Alice Dworkin, Ann Marie Kalajian, Sophie Papanikolaou, Camille Trentacoste, Dan DePasquale, and all the others at Prentice Hall who have had a hand in the project.

Mastering Mechanics I; USING MATLAB 5

1

Vector Conversion

1.1 Introduction

Vector manipulation is the basis of statics and statics in turn is the basis of engineering mechanics. These first MATLAB functions convert vectors presented in the two standard directional notation formats: rise versus run, and arbitrary angle, to x-axis and y-axis decomposition see figures 1.1 through 1.3. This conversion to a standard vector format is necessary so that rigid body equilibrium routines can be written in a more generic fashion. These conversion routines are written as function M-files rather than script M-files. Creating function files expands and customizes MATLAB.

1.2 Problem

Given a vector in either the form of rise versus run or with an arbitrary angle format, find the x-axis and y-axis decomposition. Be able to convert from this format back to the arbitrary angle formats, either degree or radian measure.

Figure 1.1   Vectors presented in rise versus run format.

Figure 1.2   Vectors presented in arbitrary angle format.

Figure 1.3   Vectors in x-axis and y-axis decomposition.

1.3   Theory

The rise versus run convention can easily be converted to and from arbitrary angle convention using the equation:

$$\tan\theta = \frac{\text{rise}}{\text{run}}$$

Eq. 1.1

Be careful when applying this equation to stay with the convention used in this book that measures angles in the positive sense, counterclockwise from the right horizontal axis. Once the vector is in arbitrary angle notation both problems can be solved by applying the equations:

$$x = m\cos\theta$$
$$y = m\sin\theta$$

Eq. 1.2

Where m is the magnitude of the original vector.

Be careful when using this function to enter a run toward the left as negative, just as a rise going downward must also be entered as negative. Care must also be taken to use the proper function depending upon the angular units, either radians or degrees.

1.4   Template

deg2xy.m

```
function [vector]=deg2xy(inputs)
%DEG2XY Converts vectors in degree angles to standard form.
%
%    [VECTOR]=DEG2XY (INPUTS) Routine takes any number of force
%    vectors described by their angle and magnitude along with an optional
%    set of coordinates and returns the force vector in standard form.
%
%    INPUTS: [ANGLE,MAG,XCOR,YCOR]
%
%    See also DIST2X, DIST2Y, DISTLOAD, RAD2XY, RISE2XY, XY2DEG, XY2RAD.

%    Details are to be found in Mastering Mechanics I, Douglas W. Hull,
%    Prentice Hall, 1999

%    Douglas W. Hull, 1999
%    Copyright (c) 1999 by Prentice Hall
%    Version 1.00

[r,c]=size(inputs);

if c==4
  ycor=inputs(:,4); % pull coordinates out of input matrix
  xcor=inputs(:,3); % pull coordinates out of input matrix
end

mag=inputs(:,2); % pull magnitudes out of input matrix
angle=inputs(:,1); % pull angles out of input matrix

if c==2 % if coordinates were not given
  xcor=zeros(r,1); ycor=zeros(r,1); % coordinates default to origin
end

angle=DR(angle); % convert to radians

xmag=cos(angle).*mag; % equation 1.2
ymag=sin(angle).*mag; % equation 1.2

vector=[xmag,ymag,xcor,ycor]; % reassemble complete answer
```

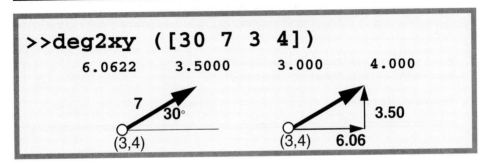

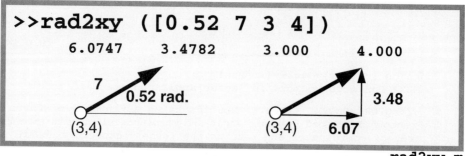

```
>>rad2xy ([0.52 7 3 4])
   6.0747      3.4782      3.000      4.000
```

rad2xy.m

```
function [vector]=rad2xy(inputs)
%RAD2XY Converts vectors in radian angles to standard form.
%    RAD2XY (INPUTS) Routine takes any number of force vectors described by
%    their angle and magnitude along with an optional set of coordinates and
%    returns the force vector in standard form.
%
%    INPUTS: [ANGLE,MAG,XCOR,YCOR]
%
%    See also DEG2XY, DIST2X, DIST2Y, DISTLOAD, RISE2XY, XY2DEG, XY2RAD.

%    Details are to be found in Mastering Mechanics I, Douglas W. Hull,
%    Prentice Hall, 1999

%    Douglas W. Hull, 1999
%    Copyright (c) 1999 by Prentice Hall
%    Version 1.00


[r,c]=size(inputs);

if c==4  % if coordinates were given
  ycor=inputs(:,4); % pull coordinates out of input matrix
  xcor=inputs(:,3); % pull coordinates out of input matrix
end

mag=inputs(:,2); % pull magnitudes out of input matrix
angle=inputs(:,1); % pull angles out of input matrix

if c==2 % if coordinates were not given
  xcor=zeros(r,1); ycor=zeros(r,1); % coordinates default to origin
end

xvec=cos(angle).*mag; % equation 1.2
yvec=sin(angle).*mag; % equation 1.2

vector=[xvec,yvec,xcor,ycor]; % reassemble complete answer
```

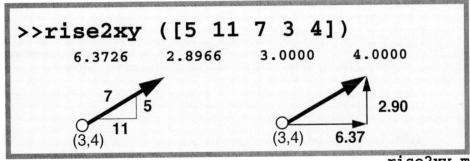

```
function [vector]=rise2xy(inputs)
%RISE2XY Converts vectors in rise-run format to standard form.
%   RISE2XY (INPUTS) Routine takes any number of force vectors described by
%   their angle and magnitude along with an optional set of coordinates and
%   returns the force vector in standard form.  If the coordinates are not
%   specified, they will default to the origin.
%
%   INPUTS: [RISE,RUN,MAG,XCOR,YCOR]
%
%   See also DEG2XY, DIST2X, DIST2Y, DISTLOAD, RAD2XY, XY2DEG, XY2RAD.

%   Details are to be found in Mastering Mechanics I, Douglas W. Hull,
%   Prentice Hall, 1999

%   Douglas W. Hull, 1999
%   Copyright (c) 1999 by Prentice Hall
%   Version 1.00


[r,c]=size(inputs);

if c==5 % if coordinates were given
   ycor=inputs(:,5); % pull coordinates out of input matrix
   xcor=inputs(:,4); % pull coordinates out of input matrix
end

mag=inputs(:,3); % pull magnitudes out of input matrix
run=inputs(:,2); % pull run out of input matrix
rise=inputs(:,1); % pull rise out of input matrix

if c==3 % if coordinates were not given
   xcor=zeros(r,1); ycor=zeros(r,1); % coordinates default to origin
end

angle=atan2(rise,run); % convert to radians

xvec=cos(angle).*mag; % equation 1.2
yvec=sin(angle).*mag; % equation 1.2

vector=[xvec,yvec,xcor,ycor]; % reassemble complete answer
```

```
ifunction [radians]=DR(degrees)
%DR Changes a matrix of degree measure to a matrix of radian measure.
%    DR(X) is the radian equivalent of the elements of X.
%
%    See also RD.
%

%    Details are to be found in Mastering Mechanics I, Douglas W. Hull,
%    Prentice Hall, 1999

%    Douglas W. Hull, 1999
%    Copyright (c) 1999 by Prentice Hall
%    Version 1.00

radians=degrees*pi/180;
```

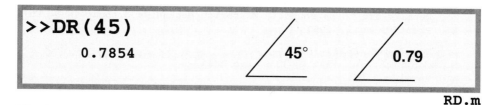

```
>>DR(45)
    0.7854                    45°        0.79
```

```
function [degrees]=RD(radians)
%RD Changes a matrix of radian measure to a matrix of degree measure.
%
%    RD(X) is the radian equivalent of the elements of X.
%
%    See also DR.

%    Details are to be found in Mastering Mechanics I, Douglas W. Hull,
%    Prentice Hall, 1998

%    Douglas W. Hull, 1998
%    Copyright (c) 1998-99 by Prentice Hall
%    Version 1.00

degrees=radians*180/pi;
```

> >>RD(0.7854)
>
> 45.0001   0.79   45°

```
function [vector]=xy2deg(inputs)
%XY2DEG Converts vectors in standard form to degree angle form.
%   XY2DEG (INPUTS) Routine takes any number of force vectors in standard form
%    and returns them in degree angle format.
%
%    INPUTS: [XMAG,YMAG,XCOR,YCOR]
%    OUTPUTS: [ANGLE,MAG,XCOR,YCOR]
%
%    See also DEG2XY, RAD2XY, RISE2XY, XY2RAD.
%

%    Details are to be found in Mastering Mechanics I, Douglas W. Hull,
%    Prentice Hall, 1999

%    Douglas W. Hull, 1999
%    Copyright (c) 1999 by Prentice Hall
%    Version 1.00


[r,c]=size(inputs);

if c==4
  ycor=inputs(:,4); % pull coordinates out of input matrix
  xcor=inputs(:,3); % pull coordinates out of input matrix
end

if c==2 % if coordinates were not given
  xcor=zeros(r,1); ycor=zeros(r,1); % coordinates default to origin
end

mag=sqrt(inputs(:,1).^2+inputs(:,2).^2); % magnitudes from inputs
angle=RD(atan2(inputs(:,2),inputs(:,1))); % angles from inputs

vector=[angle,mag,xcor,ycor]; % reassemble complete answer
```

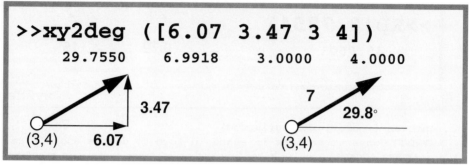

```
>>xy2deg ([6.07 3.47 3 4])
  29.7550      6.9918      3.0000      4.0000
```

xy2rad.m

```
function [vector]=xy2rad(inputs)
%XY2RAD Converts vectors in standard form to radian angle form.
%   XY2RAD (INPUTS) Routine takes any number of force vectors in standard form
%     and returns them in radian angle format.
%
%   INPUTS:  [XMAG,YMAG,XCOR,YCOR]
%   OUTPUTS: [ANGLE,MAG,XCOR,YCOR]
%
%   See also DEG2XY, RAD2XY, RISE2XY, XY2DEG.

%   Details are to be found in Mastering Mechanics I, Douglas W. Hull,
%   Prentice Hall, 1999

%   Douglas W. Hull, 1999
%   Copyright (c) 1999 by Prentice Hall
%   Version 1.00

[r,c]=size(inputs);

if c==4
  ycor=inputs(:,4); % pull coordinates out of input matrix
  xcor=inputs(:,3); % pull coordinates out of input matrix
end

if c==2 % if coordinates were not given
  xcor=zeros(r,1); ycor=zeros(r,1); % coordinates default to origin
end

mag=sqrt(inputs(:,1).^2+inputs(:,2).^2); % magnitudes from inputs
angle=atan2(inputs(:,2),inputs(:,1)); % angles from inputs

vector=[angle,mag,xcor,ycor]; % reassemble complete answer
```

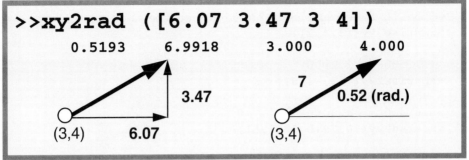

breakup.m

```
function [xmag,ymag,xcor,ycor]=breakup(vectors)
%BREAKUP Breaks a standard form force vector into its component parts.
%    [XMAG,YMAG,XCOR,YCOR]=BREAKUP(X) Subroutine that breaks a
%    multivector load matrix, X, into four column vectors representing the
%    x magnitudes, y magnitudes, x coordinates, and y coordinates.
%
%    This function is designed as a routine to be called from other
%    functions.
xmag=vectors(:,1);
ymag=vectors(:,2);
xcor=vectors(:,3);
ycor=vectors(:,4);
```

1.5   Output

Four typical vectors are presented in Figure 1.4.

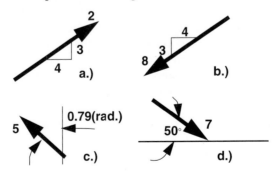

Figure 1.4   Four typical vectors in varied angle notations.

These new functions can be used at the command prompt to convert the vectors. For these examples the procedure would be:

```
>>a=rise2xy([3,4,2])
a =
    1.6000    1.2000         0         0
>>b=rise2xy([-3,-4,8])
b =
   -6.4000   -4.8000         0         0
>>c=rad2xy([0.79+0.5*pi,5])
c =
   -3.5518    3.5192         0         0
>>d=deg2xy([-50,7])
d =
    4.4995   -5.3623         0         0
```

The way this answer is read is: "The force vector *a* represents a change in the x direction of 1.6 and a 1.2 change in the y direction. This force vector originates from the origin (0,0)" The other three answers can be read in a similar manner.

Some explanation is in order for example *c*. The figure shows an angular measure of 0.79 radians yet the input angle has 0.5 π added to it. This is a way to keep with the convention of measuring from the horizontal. Since this function requires an input in radians, the 90° offset was converted to radians by hand and added. Since people think better in degrees but MATLAB works with radians, two simple conversion functions are introduced in this book. To convert from degrees to radians use the function *DR* (input degree angle). The inverse function, *RD* (input radian angle) can be used to view MATLAB output in the more comfortable degree units. With this, example *c* could have been executed in a more intuitive form.

```
>>c=rad2xy([0.79+DR(90),5])
c =
   -3.5518    3.5192         0         0
```

Do not make the mistake of saying the angle on vector *d* is 180° minus 50°. The angle is to be measured from the tail, not the head of the vector. The angle is either -50° or 310° depending on your preference.

Note also that you can convert more than one vector at a time with these routines. Each of the vectors to be converted must be entered into a large matrix, one vector per row. This command will reproduce the answers for *a* and *b*.

```
>>abinput=[3,4,2;-3,-4,8]; % combine the a and b input
vectors.
>>ab=rise2xy(abinput) % can operate on an Nx3 input vector.
ab =
    1.6000    1.2000         0         0
   -6.4000   -4.8000         0         0
```

For this book, a Standard Vector Format (SVF) was adopted.
[x magnitude, y magnitude, x coordinate, y coordinate]

Since no coordinates were given for the tail of the vectors in the above examples, the values defaulted to the origin.

Multiple vectors are put in SVF for the purpose of keeping related forces together and minimizing the number of separate arguments to a function. Although many times much of the data is not used in the function call, this method is more efficient in terms of effort expended by the user. The matrix format is being used not because matrix operations are being performed, but because it is a means of keeping the data organized.

When these functions are used to change vectors to the standard form they will usually be accompanied by their location. These particular functions do not do anything with this location data, they simply pass it along to the answer so that future functions that make use of the location data will be satisfied. To add the positional data simply add two more columns to the input matrix representing the coordinates. Examples of how to include this positional data will be covered in the next chapter.

Often when using this book the answer to some operation will be given in SVF. However, the vector might be better understood in arbitrary angle format. For this reason the two templates *xy2rad.m* and *xy2deg.m* were created.

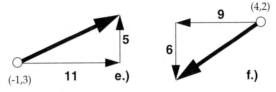

Figure 1.5   Vectors in x-axis, y-axis decomposition.

For vectors *e* and *f*, Figure 1.5 the procedure to revert back to arbitrary angle format would be:

```
>>exy=[11 5 -1 3];
>>e=xy2deg(exy)
e =
  24.4440   12.0830   -1.0000    3.0000
>>f=xy2rad([-9 -6 4 2])
f =
  -2.5536   10.8167    4.0000    2.0000
```

The way this answer is read is: "The force vector *e* is 24.4 in magnitude and it is at an angle of 12° counter clockwise from right horizontal. This force vector originates from the point (-1,3)"

1.6   Features

These routines will accept any number of input vectors as long as they are properly formatted in SVF within the input matrix. These are very useful functions used to load a rigid body in future routines.

1.7   Summary

Required argument, optional argument, [SFV]=Standard Format Vector

x=RD(*radian angle*)

　　Converts radians to degrees, input can be any size matrix.

x=DR(*degree angle)*)

　　Converts degrees to radians, input can be any size matrix.

[SFV]=Rise2xy(*[rise, run, mag, x coordinate, y coordinate]*)

　　Converts a rise versus run angle format to standard format, input can be any number of load vectors put into a larger matrix one per row.

[SFV]=Deg2xy(*[degree angle,mag, x coordinate, y coordinate]*)

　　Converts a degree angle format to standard format, input can be any number of load vectors put into a larger matrix one per row.

[SFV]=Rad2xy(*[radian angle,mag, x coordinate, y coordinate]*)

　　Converts a radian angle format to standard format, input can be any number of load vectors put into a larger matrix one per row

Standard vector format: single vector

$$\begin{bmatrix} x \text{ magnitude}, & y \text{ magnitude}, & x \text{ coordinate}, & y \text{ coordinate} \end{bmatrix}$$

Standard vector format: multiple vector

$$\begin{bmatrix} x_1\text{magnitude} & y_1\text{magnitude} & x_1\text{coordinate} & y_1\text{coordinate} \\ x_2\text{magnitude} & y_2\text{magnitude} & x_2\text{coordinate} & y_2\text{coordinate} \\ x_3\text{magnitude} & y_3\text{magnitude} & x_3\text{coordinate} & y_3\text{coordinate} \end{bmatrix}$$

2

One Point Rigid Body Equilibrium

2.1 Introduction

Often engineering mechanics problems begin with solving for reaction forces applied at a single point on a rigid body that is subjected to several known forces. The routine in this chapter will solve a surprisingly large number of problems of this nature.

2.2 Problem

Given a rigid body with any combination of coplanar point loads and moments, at a single arbitrary point, find the:
- Reaction forces in the x and y directions
- Reaction moment

Consider what this means. If a rigid body is being acted upon by a set of forces, the reaction forces are different depending upon the point of application. This routine can find the reaction forces no matter where that single point of application is located.

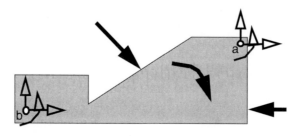

Figure 2.1   Rigid body acted upon by known set of two forces and a
                  moment. Two possible places for reaction forces are shown as
                  points *a* and *b*.

Figure 2.1 shows a typical rigid body. If the equalizing reaction forces and moment were applied at point *a* they would be different than if they were applied at point *b*. Either point could maintain equilibrium, but different moments would be required.

2.3   Theory

For static equilibrium to be achieved, all the forces in every direction must sum to zero and all the moments about every point must also sum to zero. These conditions are met when Equation Set 2.1 is true.

$$\mathbf{A}F_x = 0$$
$$\hat{\mathbf{A}}F_y = 0$$
$$\hat{\mathbf{A}}M = 0$$

Eq. 2.1

Now it is easy to see why the routines of chapter 1 were designed to break vectors into their x-axis and y-axis components. Once broken into these orthogonal components, the summation of forces is simple.

When summing moments by hand, the math is simplified by summing about a point that is on the line of action of an unknown vector or two. However, MATLAB finds it simpler to always sum the moments about the origin, rather than find a place where the math is simpler.

2.4   Template

```
function [force,moment]=reaction(vectors,coords,couples)
%REACTION Finds the reaction force and moment needed to balance a force.
%    [FORCE,MOMENT]=REACTION(AF,[X Y], COUPLES) Given an Applied Force matrix
%    in standard multi vector format and the coordinates of the unknown
%    reaction force, the routine will return the reaction force and moment
%    at that point. If no couple is specified, it defaults to zero.
%
%    See also ONEVECTOR, SUMFORCE, SUMMOMENT, THREEVECTOR, TWOVECTOR.

%    Details are to be found in Mastering Mechanics I, Douglas W. Hull,
%    Prentice Hall, 1999

%    Douglas W. Hull, 1999
%    Copyright (c) 1999 by Prentice Hall
%    Version 1.00

if nargin==2 % if couples not given
  couples=0; % couples default to zero
end;

[force,unusedcouple]=sumforce(vectors); % find sumation of vectors
moment=-(sum(couples)+summoment(vectors,coords)); % find sum of moments
force=-force; % since it is a reaction force must be negated
force(3)=coords(1); % move vector to proper spot
force(4)=coords(2); % move vector to proper spot
```

```
function [resultant,couple]=sumforce(vectors)
%SUMFORCE Sums a set of vectors into one force vector and a couple.
%    [FORCE, COUPLE]=SUMFORCE (VECTORS)  Given a set of known vectors in
%    standard multi vector format the routine will return the sum of those
%    vectors as a single vector acting through a point so that there is no
%    couple needed to balance the original.  If such a vector placement is
%    not possible, a non-zero value for the couple will be returned. This
%    usually occurs due to a force couple being formed.
%
%    See also ONEVECTOR, REACTION, SUMMOMENT, THREEVECTOR, TWOVECTOR.

%    Details are to be found in Mastering Mechanics I, Douglas W. Hull,
```

```
>>af=[deg2xy([45,4,6,1]);5,0,12,0];
>>[f,m]=reaction(af,[0,0],3)
f=
      -7.8284    -2.8284         0          0
m=
      -17.1421
```

%   Douglas W. Hull, 1999
%   Copyright (c) 1999 by Prentice Hall
%   Version 1.00

```
[xmag,ymag,xcor,ycor]=breakup(vectors); % call subroutine
couple=0; % set couple to zero
xres=sum(xmag); % x resultant
yres=sum(ymag); % y resultant

if xres==0 % if no x resultant
   ycen=0;% move resultant onto x axis
   couple=couple+sum(xmag.*(-ycor)); % check for a couple
else % there is an x resultant
   ycen=sum(xmag.*ycor)/xres; % move x res to maintain equal moment
end

if yres==0 % if no y resultant
   xcen=0; % move resultant onto y axis
   couple=couple+sum(ymag.*xcor); % check for a couple
else % there is a y resultant
   xcen=sum(ymag.*xcor)/yres; % move y res to maintain equal moment
end

resultant=[xres,yres,xcen,ycen]; % reassemble resultant matrix
```

```
>>af =[-3 0 5 3;-3 0 5 5;0 -5 6 4];
>>[f,m]=sumforce (af)
f =
            -6          -5          6          4
m =
```

summoment.m

```
function [moment]=summoment(vectors, coords)
%SUMMOMENT Solves for the moment caused by a set of forces.
%    [moment]=SUMMOMENT(VECTORS, [X,Y])  Given a set of known force
%    vectors in standard multi vector format, the routine will return the
%    resultant moment as seen from the coordinates passed in with the known
%    force vectors.  If no coordinates are specified then the moment is
%    calculated about the origin.  Right hand rule for sign convention.
%
%    See also ONEVECTOR, REACTION, SUMFORCE, THREEVECTOR, TWOVECTOR.

%    Details are to be found in Mastering Mechanics I, Douglas W. Hull,
%    Prentice Hall, 1999

%    Douglas W. Hull, 1999
%    Copyright (c) 1999 by Prentice Hall
%    Version 1.00

if nargin==1 % if coordinates are not included
  xpos=0; % coordinates default to zero
  ypos=0; % coordinates default to zero
else
  xpos=coords(1);
  ypos=coords(2);
end

[xmag,ymag,xcor,ycor]=breakup(vectors); % call subroutine
xmoment=sum(xmag.*(ypos-ycor)); % X forces times moment arm
ymoment=sum(ymag.*(xcor-xpos)); % Y forces times moment arm
moment=xmoment+ymoment; % sum both moments
```

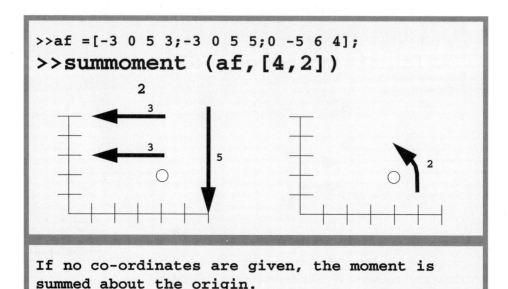

```
>>af =[-3 0 5 3;-3 0 5 5;0 -5 6 4];
>>summoment (af,[4,2])
```

If no co-ordinates are given, the moment is summed about the origin.

showvect.m

```
function []=showvect(vectors)
%SHOWVECT Draws a simple diagram showing the input vectors.
%   SHOWVECT(vectors) Shows all the input vectors on the same coordinate axis.
%   Heads of vectors are designated with an "X" while tails are marked with "O".
%
%   Since the input vectors can be any set of vectors in standard format it is
%   possible to combine the input and output of solving functions to look at the
%   relationship between the two.  The simplest way to do this would be
%   SHOWVECT ([input; output])
%
%   Use AXIS ('equal') to scale the drawing properly, may cause the vectors to
%   run off the edge of the plot. If the all of the vectors do not appear, run
%   EXPANDAXIS.
%
%   See also EXPANDAXIS, SHOWCIRC, SHOWRECT, SHOWX, SHOWY, TITLEBLOCK.

%   Details are to be found in Mastering Mechanics I, Douglas W. Hull,
%    %   Prentice Hall, 1998


%   Douglas W. Hull, 1998
%   Copyright (c) 1998-99 by Prentice Hall
%   Version 1.00
```

```
[xmag,ymag,xcor,ycor]=breakup(vectors);
xmin=min(min([xcor,xcor+xmag])); % leftmost edge
xmax=max(max([xcor,xcor+xmag])); % rightmost edge
ymin=min(min([ycor,ycor+ymag])); % lower edge
ymax=max(max([ycor,ycor+ymag])); % upper edge
xmar=max([abs(xmax-xmin)*.2,1]); % margin of 20% of width
ymar=max([abs(ymax-ymin)*.2,1]); % margin of 20% of height
xmin=xmin-xmar; % add a margin around plot
xmax=xmax+xmar;% add a margin around plot
ymin=ymin-ymar;% add a margin around plot
ymax=ymax+ymar;% add a margin around plot
clf %clear figure
hold on % stops automatic clearing of plot
for i=1:length(xmag) % do once for each vector to be ploted
  xhead = xcor(i)+xmag(i);
  xtail = xcor(i);
  yhead = ycor(i)+ymag(i);
  ytail = ycor(i);
  plot(xtail,ytail,'ro')
  plot([xtail,xhead],[ytail,yhead],'r-')
  plot(xhead,yhead,'rx')
end
hold off % starts automatic clearing of plot
axis ([xmin xmax ymin ymax]) % sets scale
showx; showy
```

2.5 Output

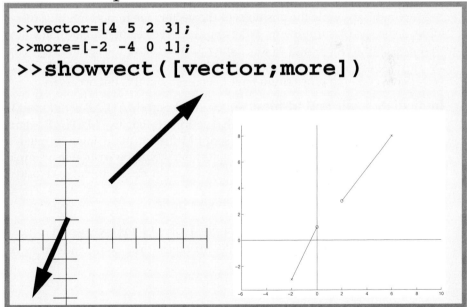

19

A typical rigid body problem is shown in Figure 2.2

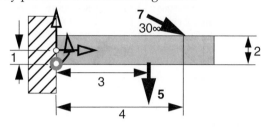

Figure 2.2   Typical one point equilibrium rigid body.

To solve this problem first the applied force matrix must be constructed using the standard multiple vector format. Then the coordinates of the resistance point are given. The result that is returned is force and moment reactions.

Creating an applied force matrix (af): When doing rigid body equilibrium problems the set of all known forces acting upon the body must be gathered. For this problem the origin was arbitrarily chosen to be at the lower left corner of the rigid body. For the body in Figure 2.2 the applied force matrix could be created as follows:

```
>>af=[deg2xy[-30,7,4,2]);0,-5,3,0]
af=
    6.0622   -3.5000    4.0000    2.0000
         0   -5.0000    3.0000         0
```

In this example the positions of the vectors are important, thus they were input as the last two arguments in the applied force vectors rather than letting the values default to the origin as in Chapter 1. Also, since the second applied force vector was already in x and y format, the values were simply entered. One of the vector conversion routines could have been used but it is not necessary.

Immediate mode and M-files: For the most part, MATLAB is not used in the immediate mode as shown above. That is to say that the MATLAB command prompt is generally used only to call script M-files or to do some simple testing. When doing a problem such as this, a script file would be written in a simple text editor. All of the code in this book can be found at the Mathworks' World Wide Web site.

ftp://ftp.mathworks.com/pub/books/hull/

Creating the M-file: Now that the applied force vector is created it can be used as an argument in this chapter's M-file, *reaction.m*. The commands to solve this problem would be:

```
af=[deg2xy([(-30),7,4,2]);0,-5,3,0];
[RForce, RMoment]=reaction (af, [0,1]);
```

That is it! Once the M-file is entered and executed, the answers can be displayed by typing the name of the variables that you want to know, *RForce* for the reaction force or *RMoment* for the reaction moment. These variable names can be typed at the command prompt, or they can be statements added at the end of the M-file so that they automatically display after the calculations are done. Use whichever method suits your needs.

>>RForce
  -6.0622   8.5000   0   1.0000
>>RMoment
  35.0622

The force vector is returned in standard format so that it may then be used as the input of a future function. The moment is returned using the right hand rule as the sign convention.

When creating applied force vectors they should be checked for accuracy, it is very easy to make mistakes if care is not taken. To simplify the checking process an M-file has been created to give a more visual representation of the vectors. The vectors are shown to scale, see Figure 2.3 with the head of the vector represented by an "X" and the tail by an "O." Use this M-file, *showvect.m* as shown:

>>showvect(af)

This function will display any set of vectors passed to it in multiple vector format. To check the answer by viewing it along with the applied force vector simply create a new multiple vector matrix for the input.

>>showvect([af;RForce])

This command concatenates the *RForce* vector as a new row at the end of the *af* matrix. This new matrix is not stored anywhere, it is created on the fly and must be recreated if it is to be used for other purposes.

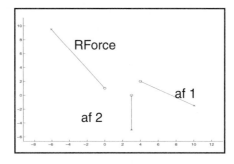

Figure 2.3   MATLAB screen shot of *showvect.m*, labels added.

2.6   Features

The function *reaction.m* has many uses if creatively applied to problems. Although the example presented here did not use a couple in the input arguments, that feature is supported. Future examples will include the use of that feature. Any number of applied force vectors may be applied to the rigid body and the function will work.

2.7   Summary

Required argument, optional argument, [SFV]=Standard Format Vector
[SFV, Moment]=Reaction(*af, X coordinate, Y coordinate,* Couple)
   Given any matrix of vectors that represent the applied forces on a rigid body and the X and Y coordinates of the reaction force, the reaction force and moment will be returned. An optional set of couples can be given as one of the input arguments. The function returns two inputs, it should be called in the above manner so that both of the inputs are saved in variables, even if only one will be used later.
ShowVect(*Vectors*)
   Given any set of vectors, a simple-to-scale graphical representation of them will be drawn. "O" represents the tail of the vector, "X" represents the head. This function is useful for checking input forces for accuracy. When the answer to a problem is found, it can be displayed with the applied forces by concatenating it to the end of the applied force vector.
   ```
   >>ShowVect([af; Answer])
   ```
      or
   ```
   >>Together = [af; Answer];
   >>ShowVect(Together)
   ```

3

Three Vector Rigid Body Equilibrium

3.1   Introduction

Often in engineering mechanics problems a rigid body is subjected to several known forces while being held in equilibrium by three forces of known placement and direction, but unknown magnitude. The routine in this chapter will solve these problems for the unknown magnitudes.

3.2   Problem

Given a rigid body with any combination of coplanar point loads and moments, find the *magnitude* of the three reaction vectors that are of
  • Known direction
  • Known placement

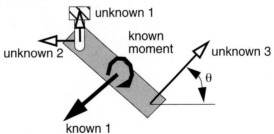

Figure 3.1   Rigid body acted upon by a known force and moment. The pin force can be thought of as two orthogonal forces of known direction and unknown magnitude and the cable force as a third force of known direction only.

The pin force in Figure 3.1 is actually of unknown magnitude and unknown direction, two unknown quantities in total. However, the pin force can be more productively thought of as two forces in the X and Y directions with unknown magnitudes, again only two unknowns. These two orthogonal forces just happen to be applied at the same point. Breaking the problem into three unknown magnitudes allows the use of the generic solving algorithm, *ThreeVec.m.*

3.3   Theory

This template is the first example of solving simultaneous linear equations with MATLAB. The simultaneous equations being solved are the same as the ones set out in Equation Set 2.1. Figure 3.3 illustrates the most generic case solvable.

$$\sum F_x = 0$$
$$\sum F_y = 0$$
$$\sum M = 0$$

Eq. 3.1

These equations ensure static equilibrium. The equations can be written for the specific subset of cases as follows:

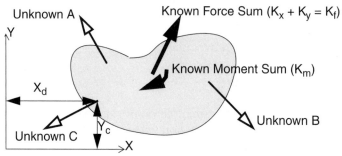

Figure 3.2   Most generic case solvable by this algorithm: Three forces of unknown magnitude but known orientation, restraining a body under the effect of known forces and moments.

$$-K_x = a\cos\theta_a + b\cos\theta_b + c\cos\theta_c$$
$$-K_y = a\sin\theta_a + b\sin\theta_b + c\sin\theta_c$$
$$-(K_m + M_{kf}) = a(X_a\sin\theta_a - Y_a\cos\theta_a) + b(X_b\sin\theta_b - Y_b\cos\theta_b) + c(X_c\sin\theta_c - Y_c\cos\theta_c)$$

Eq. 3.2

24

These three equations can be solved for.

Symbol	Meaning
a, b, c	Unknown magnitudes of reaction forces
$X_{a,b,c}$, $Y_{a,b,c}$	Known positions of reaction forces
$\theta_{a,b,c}$	Known angles of reaction forces
K_x, K_y	Summation of known forces
K_m	Summation of known moments
M_{kf}	Summation of moments caused by known forces

Table 3.1   Symbols and meanings for Equation 3.2

These can then be rearranged into matrix form so that the three unknowns can be easily found from the three equations.

$$\begin{bmatrix} \cos\theta_a & \cos\theta_b & \cos\theta_c \\ \sin\theta_a & \sin\theta_b & \sin\theta_c \\ (X_a\sin\theta_a - Y_a\cos\theta_a) & (X_b\sin\theta_b - Y_b\cos\theta_b) & (X_c\sin\theta_c - Y_c\cos\theta_c) \end{bmatrix} \begin{bmatrix} a \\ b \\ c \end{bmatrix} = \begin{bmatrix} -K_x \\ -K_y \\ -(K_m + M_{kf}) \end{bmatrix}$$

Eq. 3.3

Once in this form the three unknown magnitudes can easily be solved for by inverting the first matrix and multiplying both sides of the equation by it.

3.4   Template

threevector.m

```
function [reactions]=threevector(knowns, unknowns, couples)
%THREEVECTOR Solves for three force vectors of known direction only.
%   THREEVECTOR (KNOWNS, UNKNOWNS, COUPLES)  Routine takes a rigid body
%   acted upon by a set of known load vectors and balanced by a set of three
%   forces of known direction and unknown magnitude, and solves for the
%   previously unknown magnitudes.   The answer is returned in standard
%   multivector format. Angles are to be given in radians.
%
%   KNOWNS matrix is in standard format
%   UNKNOWNS: [ANGLE1, X1, Y1; ANGLE2, X2, Y2; ANGLE3, X3, Y3];
%   COUPLES matrix is optional [COUPLE1, COUPLE2, COUPLE3 ...];
```

```
%
%    See also ONEVECTOR, REACTION, SUMFORCE, SUMMOMENT, TWOVECTOR.

%    Details are to be found in Mastering Mechanics I, Douglas W. Hull,
%    Prentice Hall, 1999

%    Douglas W. Hull, 1999
%    Copyright (c) 1999 by Prentice Hall
%    Version 1.00

if nargin==2 couples=0; end % couples defaults to zero

coef(1,:)=cos(unknowns(:,1)');
coef(2,:)=sin(unknowns(:,1)');
coef(3,:)=unknowns(:,2)'.*coef(2,:)-unknowns(:,3)'.*coef(1,:);

answ(1,:)=(-1)*sum(knowns(:,1));
answ(2,:)=(-1)*sum(knowns(:,2));
answ(3,:)=(-1)*(summoment(knowns)+couples);

valu=inv(coef)*answ; % solving a system of three equations three unknowns

reactions(:,1)=coef(1,:)'.*valu;
reactions(:,2)=coef(2,:)'.*valu;
reactions(:,3:4)=unknowns(:,2:3);
```

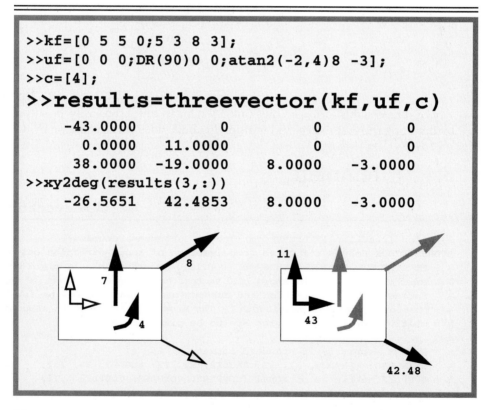

```
>>kf=[0  5  5  0;5  3  8  3];
>>uf=[0  0  0;DR(90)0  0;atan2(-2,4)8  -3];
>>c=[4];
>>results=threevector(kf,uf,c)
    -43.0000           0          0          0
      0.0000     11.0000          0          0
     38.0000    -19.0000     8.0000    -3.0000
>>xy2deg(results(3,:))
    -26.5651     42.4853     8.0000    -3.0000
```

3.5   Output

A typical problem that can be solved by this algorithm is shown in Figure 3.3.

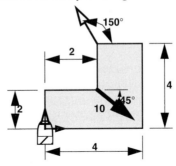

Figure 3.3   Typical rigid body that is restrained by three forces of known direction and unknown magnitude.

The use of this template requires two arguments, the applied force matrix, and matrix of position and direction of the unknown forces. An input matrix of known moments is optional.

Creating the unknowns matrix: The form of this unknowns matrix is

$$\begin{bmatrix} \theta_A \ X_A \ Y_A \\ \theta_B \ X_B \ Y_B \\ \theta_C \ X_C \ Y_C \end{bmatrix} = \begin{bmatrix} 0 & 0 & 0 \\ DR(90) & 0 & 0 \\ DR(150) & 2 & 4 \end{bmatrix} \qquad \text{Eq. 3.4}$$

This equation shows how to apply the standard format, left, to the specific example of Figure 3.3.

Creating the applied force matrix: Creating the applied force matrix for this problem is the same as making the applied force matrix for all the templates in this book. Refer to "Creating an applied force matrix (af)" Section 2.5. For this problem, the applied force matrix would be
```
>>af=[deg2xy([(-45),10,2,2])];
```

Creating the M-file: The code needed to solve this problem is very short:

CH0301.m

```
af=[deg2xy([(-45),10,2,2])];
unknowns=[0,0,0;DR(90),0,0;DR(150),2,4];
solution=threevector(af,unknowns)
showvect([af;solution]);
```

```
solution =
   -1.5840        0         0         0
    0.0000    3.9031        0         0
   -5.4871    3.1680    2.0000    4.0000
```

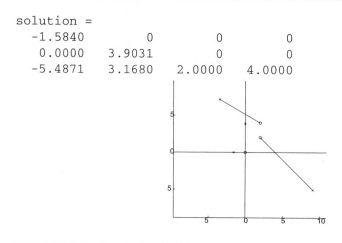

Figure 3.4   Representation of force vectors.

This short code first creates the applied force matrix, it could contain any number of known force vectors acting upon the body, see Figure 3.4. This matrix must comply with the rules set forth in "Standard vector format: multiple vector" Section 1.7

Next the unknown restraining vectors are gathered into a matrix as set forth in "Creating the unknowns matrix" Section 3.5.

The solution is called for with the *threevec.m* function. Finally the solution is visually checked with the *showvect.m* function.

3.6   Features

The function *threevector.m* has many uses if creatively applied to problems, together with *reaction.m* most rigid body statics problems can be solved. Future examples will show some of these uses.

3.7   Summary

Required argument, optional argument, [SFV] = Standard Format Vector
[SFV]=threevector(af, Unknowns, Couple)
    Given any matrix of vectors that represent the applied forces on a rigid body and the unknown matrix, shown at right. An optional couple vector can be included to account for any couples acting on the body.

$$\begin{bmatrix} \theta_1 & x_1 & y_1 \\ \theta_2 & x_2 & y_2 \\ \theta_3 & x_3 & y_3 \end{bmatrix}$$

4

Linearly Distributed Loads

4.1   Introduction

Often times forces are not applied to a body in a way that can be directly modeled as a point force. Rather than a force being applied by a cable or at a pinned connection, the force is applied as a pressure over a given area. This pressure may be the water held back by the wall of a dam or it may be the weight of a large pile of sand. In many cases this distributed load may be converted to a point load for simplicity in calculation.

4.2   Problem

Given a force stated as a linearly distributed load find the magnitude and placement of the equivalent point force, illustrated in Figure 4.1.

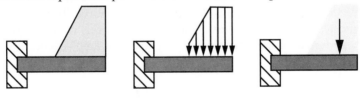

Figure 4.1   Physical situation and two schematic drawings representations.

4.3 Theory

In two–dimensional problems, the most common forms of distributed loads are constant pressure over the length of application and a pressure that varies linearly over the length of application. An example of these two types of linearly distributed forces are shown in Figure 4.2.

Figure 4.2   Examples of linearly distributed loads.

The important attributes of these types of loads are detailed and the meanings of the returned values are illustrated in Figure 4.3.

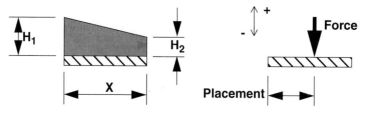

Figure 4.3   Input arguments and results returned.

With the three inputs the equivalent point load is arrived at with the help of the shape routines that are described in Chapter 8. In short the magnitude of the point force is equal to the area under the force curve. The equivalent force acts through the centroid of the area. The function for this chapter simply calls these two shape functions together since this type of problem will occur so frequently.

4.4   Template

`distload.m`

```
function [force, placement]=distload(ho,ht,x)
%DISTLOAD Converts a linearly distributed load to a point force.
%   [FORCE, PLACEMENT]=DISTLOAD(H1,H2,X) Given the two magnitudes H1 and H2
%   that a linearly distributed load varies between, and the length of
%   application, the routine will find the magnitude and location of the
%   equivalent force.
%
%   See also DEG2XY, DIST2X, DIST2Y, RAD2XY, RISE2XY, XY2DEG, XY2RAD.

%   Details are to be found in Mastering Mechanics I, Douglas W. Hull,
%   Prentice Hall, 1999

%   Douglas W. Hull, 1999
%   Copyright (c) 1999 by Prentice Hall
%   Version 1.00

if ho >= ht %left side is bigger
   force = vertrap(ho,x,ht,(ho-ht),'area'); % find area
   placement = vertrap(ho,x,ht,(ho-ht),'centX'); % find centroid
else %right side is bigger
   force = vertrap(ht,(-x),ho,(ht-ho),'area'); % find area
   placement = x + (vertrap(ht,(-x),ho,(ht-ho),'centX')); % find
centroid
end
```

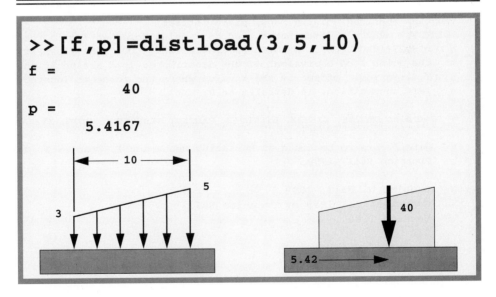

```
>>[f,p]=distload(3,5,10)
f =
          40
p =
       5.4167
```

dist2x.m

```
function [pntload]=dist2x(mags, place, offset)
%DIST2X Converts a distributed load to a point force acting in the X.
%    DIST2X([MAGNITUDE1, MAGNITUDE2],[LOCATION1, LOCATION2], OFFSET) is
%    the point load equivalent of the distributed load acting in the
%    X direction.  Offset is the Y value where the force is located, if
%    left unspecified, it defaults to 0;
%
%  See also DEG2XY, DIST2Y, DISTLOAD, RAD2XY, RISE2XY, XY2DEG, XY2RAD.

%    Details are to be found in Mastering Mechanics I, Douglas W. Hull,
%    Prentice Hall, 1999

%    Douglas W. Hull, 1999
%    Copyright (c) 1999 by Prentice Hall
%    Version 1.00

if nargin<3
  offset=0;
end

for i=1:rows(mags)
  [F, P]=distload(mags(i,1),mags(i,2),place(i,2)-place(i,1));
  pntload(i,:)=[F 0 offse place(i,1)+P];
end
```

dist2y.m

```
function [pntload]=dist2y(mags, place, offset)
%DIST2Y Converts a distributed load to a point force acting in the Y.
%    DIST2Y([MAGNITUDE1, MAGNITUDE2],[LOCATION1, LOCATION2], OFFSET) is
%    the point load equivalent of the distributed load acting in the
%    Y direction.  Offset is the X value where the force is located, if
%    left unspecified, it defaults to 0;
%
%  See also DEG2XY, DIST2X, DISTLOAD, RAD2XY, RISE2XY, XY2DEG, XY2RAD.

%    Details are to be found in Mastering Mechanics I, Douglas W. Hull,
%    Prentice Hall, 1999

%    Douglas W. Hull, 1999
%    Copyright (c) 1999 by Prentice Hall
%    Version 1.00

if nargin<3
  offset=0;
end
```

```
for i=1:rows(mags)
  [F, P]=distload(mags(i,1),mags(i,2),place(i,2)-place(i,1));
  pntload(i,:)=[0 F place(i,1)+P offset];
end
```

4.5   Output

What is the equivalent point force that can be used in the place of the distributed load of Figure 4.4.?

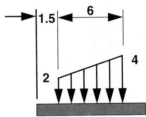

Figure 4.4   A typical distributed load.

For this routine, the three arguments simply need to be plugged into the function call

```
>>[Force, Placement]=distload(-2,-4,6)
Force =
       -18
Placement =
    3.3333
```

This routine is simple, but it is very useful. A routine that calls the *distload* function and puts the distributed load into a force vector automatically is *dist2x* or *dist2y*. These secondary functions will more likely be used than the lower level *distload*, but each has its place.

```
>>Yforce=dist2y([-2,-4],[1.5 7.5])
Yforce =
       0      18    4.8333          0
```

If the distributed load were oriented along the x axis:

```
>>Xforce=dist2x([2 4],[1.5 7.5])
Xforce =
   18.0000        0        0    4.8333
```

4.6   Features

Care must be taken to ensure that the distributed force is in the same direction across the entire length of application. For example if the force varies from -3 to 4 over the length of application, the force should be broken into two distributed loads, one from -3 to 0 in one direction and the other from 0 to 4 in the opposite direction.

Once the results are obtained in the form of force and placement some trigonometry based on the geometry of the problem may be needed to use this simplified force in another routine such as *deg2xy.m* or *reaction.m*.

Two new functions, *rows.m* and *cols.m*, were used without being formally introduced. These two are functions that are not strictly needed, their purpose is to simplify a function call from a difficult to remember syntax to an easy to remember one. Both of these new functions simply call the *size.m* function. They are merely simplifications of syntax, they add usability not functionality.

4.7   Summary

Required argument, optional argument, [SFV] = Standard Format Vector
[force, placement] = distload (h_1, h_2, x_2-x_1)

The illustration of the input and outputs are reprinted here from the original Figure 4.3.

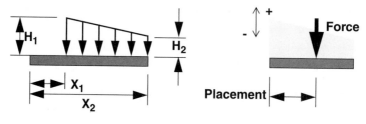

Figure 4.5   Input arguments and results returned.

sfv=dist2x ([h_1, h_2], [x_1,x_2])
Returns the point load equivalent if the distributed load is oriented along the x axis.

sfv=dist2y ([h_1, h_2], [y_1,y_2])
Returns the point load equivalent if the distributed load is oriented along the y axis.

5

Internal Truss Forces— Method of Joints

5.1   Introduction

A truss is a useful engineering abstraction because of the simplifying assumptions that allow it to be solved easily. The members of a truss are thin and contact each other only at their end points. Many real structures can be adequately modeled as trusses. Good examples of trusses are the common roof truss and the truss bridge.

5.2   Problem

With the tools presented in Chapter 3, "Three Vector Rigid Body Equilibrium", the external restraining forces on a truss can be solved for. However the internal forces are just as important and need to be solved for also. The goal of this chapter is to be able to solve for the internal forces on any statically determinant truss. To do this the previous templates will be needed, along with skills such as identifying zero force. Also needed will be some simple routines to move vectors from point to point. There also needs to be an easy means to find the opposite of a force.

5.3   Theory

The two constraining assumptions that define a structure as a truss are:
  • All applied forces are point loads applied at the joints.
  • The joints can not transmit a moment.

These assumptions ensure that each member in the truss acts as a two-force member, meaning that all forces act along the line between the two connection points. Figure 5.1 illustrates a simple truss.

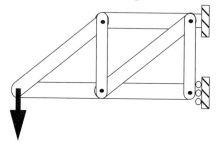

Figure 5.1   A simple truss.

The Method of Joints: If a truss is in equilibrium, then it follows that each of its joints is also in equilibrium.The procedure for solving a truss then is:
1.   Solve the truss for external reaction forces.
2.   Identify zero force members.
3.   Solve for the forces acting at each joint.
        Step one is easy using the methods of three vector rigid body equilibrium. Step two involves finding zero force members, or members that carry no load. This is not strictly necessary, but it does simplify and speed the process. Finally, at each of the joints the forces must be solved for. For this final step, *twovector.m* was designed.
        If a joint is solvable, it will have no more than two unknown forces acting on it along with any number of known forces. These unknown forces will come from truss members. Because they are two-force members the direction of the forces will be known. This leaves two unknown magnitudes and two force equations, a solvable system.
        Figure 5.2 shows the most generic case of a joint with two unknown forces that can be solved.

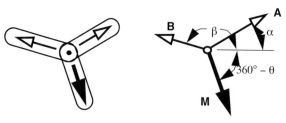

Figure 5.2   Most generic case solvable by *twovector.m*.

Because there is no moment possible at the joint, only the two force equations need to be satisfied for equilibrium.

$$\sum F_x = 0 = A\cos\alpha + B\cos\beta + M\cos\theta$$
$$\sum F_y = 0 = A\sin\alpha + B\sin\beta + M\sin\theta$$

Eq. 5.1

Rearranging this into a form easily solved for magnitudes A and B.

$$\begin{bmatrix} \cos\alpha & \cos\beta \\ \sin\alpha & \sin\beta \end{bmatrix} \begin{bmatrix} A \\ B \end{bmatrix} = \begin{bmatrix} -M\cos\theta \\ -M\sin\theta \end{bmatrix}$$

Eq. 5.2

In this situation the force vector M could be a single-force vector, or it could be the sum of any number of known force vectors acting on the joint.

In the situation where only one vector is unknown at a point, the value of the unknown force is simply the vector opposite to the sum of the known vectors. The old file *sumforce.m* should not be used for this task however. The reason is that the M-file will relocate the reaction force in an attempt to keep the moment about the origin the same for both the input and output. Also, the *onevector.m* file is superior because it checks to ensure that all the force vectors do indeed originate from the same point. This is helpful in catching errors before they occur.

5.4   Template

twovector.m

```
function [reactions]=twovec(knowns, unknowns)
%TWOVECTOR Solves for two force vectors of known direction only.
%    TWOVECTOR(KNOWNS, UNKNOWNS)  Routine takes a single point acted upon by
%    a set of known load vectors and balanced by a set of two forces of known
%    direction and unknown magnitude, and solves for the previously unknown
%    magnitudes. The answer is returned in standard multi vector format.
%    Angles are to be given in radians. This routine is particularly designed
%    for truss problems.
%
%    KNOWNS matrix is in standard format
%    UNKNOWNS matrix [ANGLE1, ANGLE2];
%
%    See also ONEVECTOR, REACTION, SUMFORCE, SUMMOMENT, THREEVECTOR.


%    Details are to be found in Mastering Mechanics I, Douglas W. Hull,
%    Prentice Hall, 1999


%    Douglas W. Hull, 1999
%    Copyright (c) 1999 by Prentice Hall
%    Version 1.00
```

```
[xmag ymag xcor ycor]=breakup(knowns);
flagx = xcor ~= mean(xcor)*ones(size(xcor));
flagy = ycor ~= mean(ycor)*ones(size(ycor));
if (flagx | flagy)
  disp ('In twovec.m all vectors must originate from the same point')
  return
end % if not all from same point
knowns=[sum(xmag) sum(ymag) xcor(1) ycor(1)];
[xmag ymag xcor ycor]=breakup(knowns);

k=sqrt(xmag^2+ymag^2);
angle=atan2(ymag, xmag);
alpha=unknowns(1);
beta=unknowns(2);

coef=[cos(alpha) cos(beta);sin(alpha) sin(beta)];
answ=-k*[cos(angle);sin(angle)];

mag=inv(coef)*answ;

reactions=[rad2xy([alpha mag(1) xcor ycor; beta mag(2) xcor ycor])];
```

```
>>knowns=[7 2 2 1; -2 3 2 1];
>>unknowns=[atan2(-3,-4), DR(180)];
>>f=twovector(knowns,unknowns)
f =
        -6.6667      -5.0000       2.0000       1.0000
         1.6667       0.0000       2.0000       1.0000
```

onevector.m

```
function [resultant]=onevector(knowns)
%ONEVECTOR Vector that is the negative of sum of forces acting at a point.
%   ONEVECTOR(KNOWNS)  Routine takes a single point acted upon by a set of
%   known force vectors and balanced by one unknown force and solves for the
%   unknown force. The answer is returned as a standard multi vector format.
%   This routine is particularly designed for truss problems.
```

```
%
%    KNOWNS matrix is in standard multi vector format.
%
%    See also REACTION, SUMFORCE, SUMMOMENT, THREEVECTOR, TWOVECTOR.

%    Details are to be found in Mastering Mechanics I, Douglas W. Hull,
%    Prentice Hall, 1999

%    Douglas W. Hull, 1999
%    Copyright (c) 1999 by Prentice Hall
%    Version 1.00

[xmag ymag xcor ycor]=breakup(knowns);
flagx = xcor ~= mean(xcor)*ones(size(xcor));
flagy = ycor ~= mean(ycor)*ones(size(ycor));
if (flagx | flagy)
   disp ('In onevec.m all vectors must originate from the same point')
   return
end % if not all from same point

resultant=[-sum(xmag) -sum(ymag) xcor(1) ycor(1)];
```

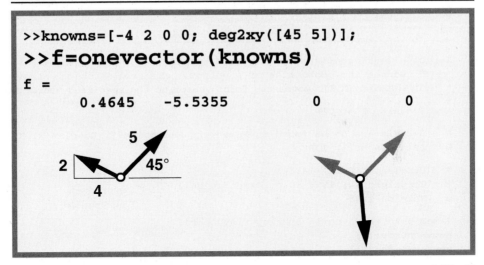

```
>>knowns=[-4 2 0 0; deg2xy([45 5])];
>>f=onevector(knowns)
f =
          0.4645    -5.5355           0           0
```

opp.m

```
function [outVector]=opp(inVector)
%OPP Returns the equal but opposite vector
%    OPP(VECTOR) returns the vector acting in the opposite direction to the
%    input vector.
%
%    See also MAG, MOVE.
```

```
%    Details are to be found in Mastering Mechanics I, Douglas W. Hull,
%    Prentice Hall, 1999

%    Douglas W. Hull, 1999
%    Copyright (c) 1999 by Prentice Hall
%    Version 1.00

[xmag ymag xcor ycor]=breakup(inVector);

outVector=[-xmag -ymag xcor ycor];
```

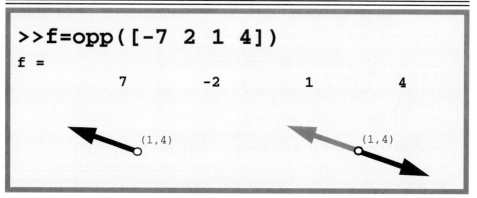

```
>>f=opp([-7 2 1 4])
f =
        7        -2        1        4
```

(1,4)   (1,4)

move.m

```
function [outVector]= move (inVector, coords)
%MOVE Changes the coordintes of a vector.
%    MOVE(VECTOR,[X,Y]) moves the force vector to the specified coordinates.
%
%    See also MAG, OPP.

%    Details are to be found in Mastering Mechanics I, Douglas W. Hull,
%    Prentice Hall, 1998

%    Douglas W. Hull, 1998
%    Copyright (c) 1998-99 by Prentice Hall
%    Version 1.00

[xmag ymag xcor ycor]= breakup (inVector);
newx=coords(1);
newy=coords(2);

xcor=ones(size(xcor))*newx;
ycor=ones(size(ycor))*newy;

outVector=[xmag ymag xcor ycor];
```

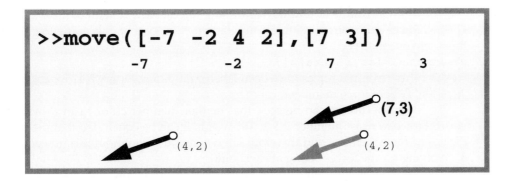

5.5   Output

A simple truss that can be solved with the algorithms presented here is shown in Figure 5.3.

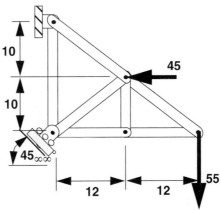

Figure 5.3   A simple truss loading.

To solve this problem many of the previously introduced tools will be needed. The best procedure is to follow the method of joints.

1.   Solve the truss for external reaction forces.

Looking at Figure 5.4 it can be seen that the *threevector.m* function should be used. This can be done by treating the entire truss as a rigid body. By doing this the reaction forces can be solved for.

```
>>af=[-45 0 12 10; 0 -55 24 0];
>>unknowns=[DR(45) 0 0; 0 0 20; DR(90) 0 20];
>>restrain=threevector(af,unknowns);
```

It is that simple to find the restraining forces. Now that all of the external applied forces and restraining forces are known, it is good bookkeeping to gather them into one matrix.

```
>>external=[af;restrain];
```

These external forces have been labeled e.f. 1–5 in Figure 5.4.

2. Identify zero force members.

By inspection it can be seen that neither member *e* nor *f* can provide a vertical force to resist member *c*. For this reason it can be concluded that member *c* is a zero-force member, and therefore safely ignored throughout this procedure.

3. Solving for the forces acting at each joint.

To solve for the internal forces on the truss, each joint must be looked at as an individual free body diagram. Then each joint must be solved independently. It is best to start at a joint that has only two unknown forces acting upon it.

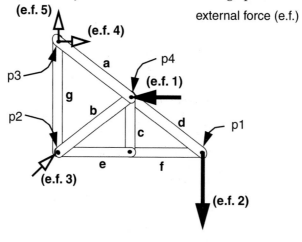

Figure 5.4   Labeled schematic drawing of truss.

```
>>point1=twovector(external(2,:),[atan2(10,-12) DR(180)]);
>>fDp1=point1(1,:);
>>fFp1=point1(2,:);
```

Point 1

Twovector.m was used to find the internal forces and store them in the matrix *point1*, then the internal forces were pulled out into their own separate force vectors. The notation can be read "Force from member d acting on point 1."

Because member c is a zero force member, members e and f can be thought of as one solid member. The internal force from member f at point 1 is the same magnitude but in the opposite direction of the internal force from member e at point 2. To move and change the sense of the force, the functions *opp.m* and *move.m* are used in conjunction.

```
>>fEp2=opp(move(fFp1,[0,0]));
```

With member e solved there are only two unknown forces acting on point 2. This means that point 2 can be solved just as point 1 was.

```
>>point2=twovector([fEp2; external(3,:)],[atan2(10,12)
DR(90)]);
>>fBp2=point2(1,:);
>>fGp2=point2(2,:);
```

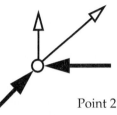

Again the internal force from member g can be inverted and moved from point 2 to the other end of the two–force member, point 3. Now point 3 has only one unknown force acting on it. This means that the remaining unknown force alone must counterbalance the sum of the known force acting there. The *onevector.m* function should be used here.

Point 2

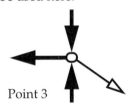

Point 3

```
>>fGp3=opp(move(fGp2,[0,20]));
>>fAp3=onevector([external(4:5,:);fGp3]);
>>fAp4=opp(move(fAp3,[12,10]));
>>fBp4=opp(move(fBp2,[12,10]));
>>fDp4=opp(move(fDp1,[12,10]));
```

This is the last step in solving for all of the internal forces of the truss. The internal forces then can be gathered into one matrix as a measure of good bookkeeping.

```
>>internal=[fDp1;fFp1;fBp2;fEp2;fGp2;fGp3;fAp3;fAp4;fBp4;fDp4];
```

For quick and easy reference the entire M-file needed to solve the truss is gathered here.

CH0401.m

```
af=[-45 0 12 10; 0 -55 24 0];
unknowns=[DR(45) 0 0; 0 0 20; DR(90) 0 20];
restrain=threevector(af,unknowns);
external=[af;restrain];

point1=twovector(external(2,:) ,[atan2(10,-12) DR(180)]);
fDp1=point1(1,:);
fFp1=point1(2,:);

fEp2=opp(move(fFp1,[0,0]));
```

```
point2=twovector([fEp2; external(3,:)],[atan2(10,12) DR(90)]);
fBp2=point2(1,:);
fGp2=point2(2,:);
fGp3=opp(move(fGp2,[0,20]));

fAp3=onevector([external(4:5,:);fGp3]);

fAp4=opp(move(fAp3,[12,10]));
fBp4=opp(move(fBp2,[12,10]));
fDp4=opp(move(fDp1,[12,10]));

internal=[fDp1;fFp1;fBp2;fEp2;fGp2;fGp3;fAp3;fAp4;fBp4;fDp4];
```

5.6 Features

These functions are useful in truss problems, but with creative use they can be applied in many other applications. The mathematics is often the same between seemingly dissimilar problems. By consulting the theory and the templates sections the compatibility of these routines and the new problems can be determined. Possibly a slight modification of the original functions can increase the useability for further applications.

5.7 Summary

Required argument, optional argument, [SFV]=Standard Format Vector

[SFV]=twovector(*knowns*, [*angle1 angle2*])
   Given any number of known vectors acting on a point and the direction of the two vectors of unknown magnitude that maintain equilibrium of the point, the magnitudes will be found. The returned value is the two restraining vectors given in SFV.

[SFV]=onevector(*knowns*)
   Given any number of known vectors acting on a point, the single vector that acts as a counterbalance will be found and returned in SFV.

[SFV]=opp(*vectors*)
   Given any number of input vectors this routine will return the vectors that are of the same magnitude but opposite direction.

[SFV]=move(*vectors*, [*New_x_coordinate New_y_coordinate*])
   Given any number of input vectors this routine will return all of the vectors with the same magnitude and direction, however the vectors will all be moved to the specified point.

6

Statics—Examples

6.1 Single Vector Entry

For each of the illustrated vectors in Figure 6.1, create a standard format vector.

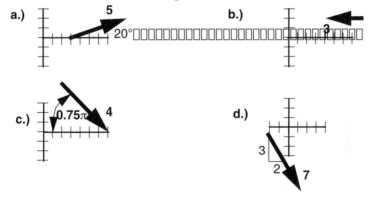

Figure 6.1   Single vectors.

```
>>a=deg2xy([20,5,3,0])
   4.6985    1.7101    3.0000         0
>>b=[-3,0,4,2]
       -3         0         4         2
>>c=rad2xy([-0.25*pi,4,7,0])
   2.8284   -2.8284    7.0000         0
```

```
>>d=rise2xy([-3,2,7,-2,-1])
   3.8829   -5.8244   -2.0000   -1.0000
```

6.2   Creating an Applied Force Matrix

For each of the illustrated free bodies in Figure 6.2, create an applied force matrix in standard multi vector format.

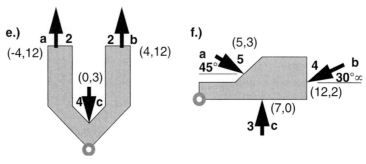

Figure 6.2   Free body diagrams.

```
>>ea=[0,2,-4,12];
>>eb=rad2xy([pi/2,2,4,12]);
>>ec=deg2xy([-90,4,0,3]);
>>e=[ea;eb;ec]
        0    2.0000   -4.0000   12.0000
   0.0000    2.0000    4.0000   12.0000
   0.0000   -4.0000         0    3.0000
>>f=[deg2xy([-45,5,5,3]);deg2xy([210,4,12,2]);0,3,7,0]
   3.5355   -3.5355    5.0000    3.0000
  -3.4641   -2.0000   12.0000    2.0000
        0    3.0000    7.0000         0
>>showvect(f)
```
        At first glance there seems to be an error in either the applied force matrix *f* or in the *showvect* representation of it shown in Figure 6.3. This is not so. Because these applied force vectors can be moved along their line of action the two pictures are indeed equivalent.

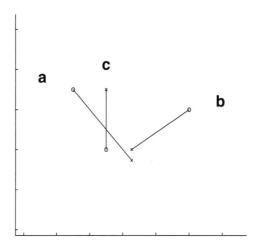

Figure 6.3   *Showvect* representation of example *f*, labels added.

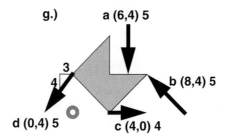

Figure 6.4   Free body diagram.

```
>>ga=[0,-5,6,4];
>>gb=deg2xy([135,5,8,4]);
>>gc=[4,0,4,0];
>>gd=rise2xy([-4,-3,5,0,4]);
>>g=[ga;gb;gc;gd]
g =
          0   -5.0000    6.0000    4.0000
    -3.5355    3.5355    8.0000    4.0000
     4.0000         0    4.0000         0
    -3.0000   -4.0000         0    4.0000
```

47

It may be instructive to look at the *showvect.m* representations of these free body diagrams. It takes a little bit of imagination to see the relationship between the illustrated diagram and the *showvect* representation. It may be helpful to sketch the rigid body onto a printout of the *showvect* picture.

6.3   Summing Force Vectors

What is the magnitude of the sum of the force vectors shown in Figure 6.5?

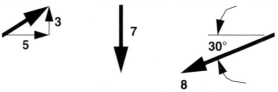

Figure 6.5   Force vectors to be summed.

```
>>vectors=[5 3 0 0;0 -7 0 0;deg2xy([180+30,8])];
>>summed=sumforce(vectors)
summed =
   -1.9282   -8.0000          0          0
```

This adds all of the force vectors together; however, it does not give the total magnitude of the force. To get the magnitude, the *mag.m* function should be used.

```
>>mag(summed)
ans =
    8.2291
```

From the last command, the magnitude of the force can be read off as 8.2291. *Mag.m* is a new function created for this text. Do a "help" for more information.

6.4   Summing Moments

What is the moment required at the center point of the bar to resist the moments caused by the two forces shown in Figure 6.6?

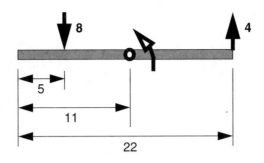

Figure 6.6   Forces causing a moment.

```
>>af=[0 -4 -6 0; 0 8 11 0];
>>moment=summoment(af)
moment =
      112
```

This number, 112, is the sum of the moments caused by the two forces. However, the moment that would resist these forces is needed, so the value must be negated. Remember that these templates use the right-hand rule convention for moment sign conventions.

6.5   Distributed Loading

What is the applied force matrix for the situations illustrated in Figure 6.7?

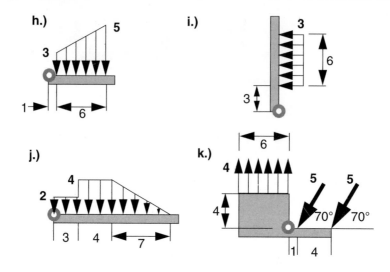

Figure 6.7   Linearly distributed loads.

```
>>h=dist2y([-3 -5],[1 7])
h =
        0 -24.0000    4.2500         0
>>i=dist2x([-3 -3],[3 9])
i =
      -18         0         0         6
>>j(1,:)=dist2y([-2 -2],[0 3]);
>>j(2,:)=dist2y([-4 -4],[3 7]);
>>j(3,:)=dist2y([-4 0],[7 14]
j =
        0  -6.0000    1.5000         0
        0 -16.0000    5.0000         0
        0 -14.0000    9.3333         0
>>k(1,:)=dist2y([4 4],[0 6],4);
>>k(2,:)=deg2xy([180+70,5,1,0]);
>>k(3,:)=deg2xy([180+70,5,5,0])
k =
        0  24.0000   -3.0000    4.0000
  -1.7101  -4.6985    1.0000         0
  -1.7101  -4.6985    5.0000         0
```

It may seem counterintuitive at first to enter the values for magnitudes of distributed loads as negative, but it is necessary to make sure that the applied force vector gives the proper sense to the load.

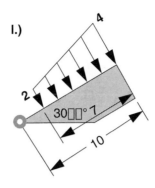

Figure 6.8   Distributed load.

This distributed load in Figure 6.8 presents a bit more of a challenge. It must be taken in several small but manageable steps. First the distributed load must be converted to a point load, then the point load must be properly oriented.

```
>>[lf,lp]=distload(2,4,7);
>>l=deg2xy([30-90,lf,(3+lp)*cos(DR(30)),(3+lp)*sin(DR(30))])
l =
   10.5000 -18.1865    5.9660    3.4444
```

Much of the above line could use some clarification. The four arguments into the function could have been made in separate declarations and then fed into the function as variable names. This improves clarity in code, but also increases its length. Regardless, justification for each of the four arguments follows.

Angle: The load is acting on a rigid body that makes a 30° angle with horizontal. The load itself, however, acts on a line of action perpendicular to the body so the 90° is subtracted off.

Force: The load equivalent to the distributed load was returned directly from the *distload.m* function. No further modification was needed.

Coordinates: The distributed load is offset by three units from the origin, and the equivalent load is further yet because it acts in the centroid of the distributed load. Once added together some trigonometry is used to find the proper location of the load in space. Also note that the *DR* function is used to convert from degrees to radians for use in the trigonometry functions.

6.6   Single Point Reaction Forces

For each of the examples shown in Figure 6.9, find the reaction forces and moment at each of the designated points.

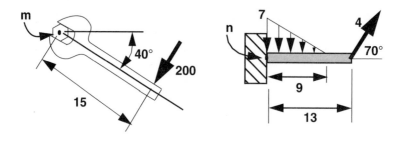

Figure 6.9   Single point reaction.

CH0601.m

```
mangle=-40-90;
mmagnitude=200;
mxcord=15*cos(DR(40));
mycord=15*sin(DR(-40));
mload=deg2xy([mangle,mmagnitude,mxcord,mycord]);
[mRforce,mRmoment]=reaction(mload,[0,0]);
```

```
>>mRforce
mRmoment =
 128.5575 153.2089        0         0
>>mRmoment
mRmoment =
     3000
```

CH0602.m

```
[naforce, naplacement]=distload(7,0,9);
na=[0,-naforce,naplacement,0];
nb=deg2xy([70,4,13,0]);
nload=[na;nb];
[nRforce, nRmoment]=reaction(nload,[0,0]);
```

```
>>nRforce
nRforce =
  -1.3681   27.7412        0         0
>>nRmoment
nRmoment =
  45.6360
```

6.7   Single Point Reaction Comparison

For this design, a choice must be made as to the location for a single restraining location. Since the force components will be the same at either location, the decision will be based on which position has the lowest moment associated with it. See Figure 6.10.

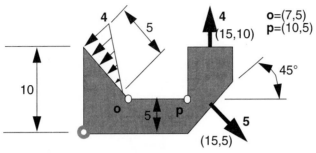

Figure 6.10   Single point reaction comparison.

CH0603.m

```
[opforce, opplacement]=distload(4,0,5);
hypot=hyp(7,5);
opax=opplacement*(7/hypot);
opay=10-opplacement*(5/hypot);
opload(1,:)=rise2xy([-7,-5,opforce,opax,opay]);
opload(2,:)=[0,4,15,10];
opload(3,:)=deg2xy([-45,5,15,5]);
[oRforce,oRmoment]=reaction(opload,[7,5]);
[pRforce,pRmoment]=reaction(opload,[10,5]);
```

When checking the forces, as expected they are equal in magnitude.
```
>>oRforce
oRforce =
    2.2768    7.6729    7.0000    5.0000
>>pRforce
pRforce =
    2.2768    7.6729   10.0000    5.0000
```
When checking the moments though, point *o* is shown to be the better choice based on the criteria of lowest moment.
```
>>oRmoment
```

```
oRmoment =
 -73.0723
>>pRmoment
pRmoment =
 -96.0909
```

6.8   Three Force Reaction

In the following two problems, there are three reaction forces of unknown magnitude but of known direction. Find these forces and put them in standard format. In example q the beam is supported by a cable on one end and a pin joint on the other. What are the forces that hold this body in static equilibrium?

q.)

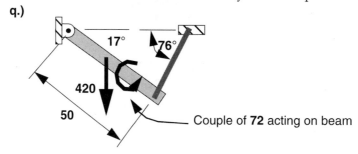

Couple of **72** acting on beam

Figure 6.11   Three point equilibrium.

At first glance this *threevector.m* routine does not seem appropriate since it applies to situations where the three unknown are magnitudes. The cable supplies a force of known direction and unknown magnitude, so it is usable. However, the pin force seems to be of unknown force and direction. This of course is true, but this same force can be thought of as two forces of known direction, x and y, and unknown magnitude. With this idea the problem can be solved in the routine.

CH0604.m

```
qcordx=25*cos(DR(-17));
qcordy=25*sin(DR(-17));
qload=[0,-420,qcordx,qcordy];
qunknowns(1,:)=[0,0,0];
qunknowns(2,:)=[DR(90),0,0];
qunknowns(3,:)=[DR(76),50*cos(DR(-17)),50*sin(DR(-17))];
qcouple=72;
q=threevector(qload,qunknowns,qcouple)
```

```
q =
   -48.3016          0          0          0
         0   226.2731          0          0
    48.3016   193.7269    47.8152   -14.6186
```

Figure 6.12 shows a problem that is conceptually the same and is easily solved as shown.

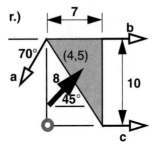

Figure 6.12   Three point equilibrium.

CH0605.m

```
rload=deg2xy([45,8,4,5]);
runknowns=[DR(180+70),0,10;0,7,10;0,7,0];
r=threevector(rload,runknowns);
```

```
r =
   -2.0589   -5.6569         0   10.0000
    1.4932         0    7.0000   10.0000
   -5.0912         0    7.0000         0
```

6.9   Truss Internal Forces

What are the internal forces in all members of this truss as illustrated in Figure 6.13?

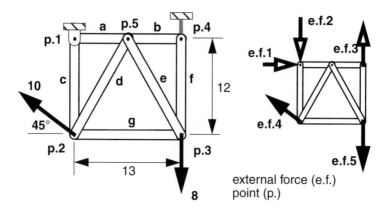

Figure 6.13   Loaded truss.

CH0606.m

```
af=[deg2xy([45 10 0 0]);0 -8 13 0];
restrain=threevector(af,[0 0 12;DR(-90) 0 12;DR(90) 13 12]);
external=[restrain; af];
p1=twovector(external(1:2,:),[0 DR(90)]);
fAp1=p1(1,:);
fCp1=p1(2,:);
fCp2=opp(move(fCp1,[0,0]));
p2=twovector([external(4,:);fCp2],[atan2(6.5,12),0]);
fDp2=p2(1,:);
fGp2=p2(2,:);
fGp3=opp(move(fGp2,[13,0]));
p3=twovector([external(5,:);fGp3],[atan2(12,-6.5),DR(90)]);
fEp3=p3(1,:);
fFp3=p3(2,:);
fFp4=opp(move(fFp3,[13,12]));
fBp4=onevector([external(3,:);fFp4]);
fAp5=opp(move(fAp1,[6.5,12]));
fBp5=opp(move(fBp4,[6.5,12]));
fDp5=opp(move(fDp2,[6.5,12]));
fEp5=opp(move(fEp3,[6.5,12]));
ifAD=[fAp1;fAp5;fBp5;fBp4;fCp1;fCp2;fDp2;fDp5];
ifEF=[fEp3;fEp5;fFp3;fFp4;fGp2;fGp3];
internal=[ifAD;ifEF];
```

Inspection of the results shows that member b is a two–force member. Noticing this before solving would have made the solution that much shorter. Even though that simplifying observation was not made in time, the solution method still worked. An easy way to help verify the results are to run the internal forces through the *sumforce.m* function all at once. The sum of the internal forces should be zero. This is also true of the external forces.

It may be instructive to view the results with the *showvect.m* function. Try viewing the internal forces. Notice how the force vectors are parallel to the member itself. This should always be true with two-force members.

7

Geometry of Primitive Shapes

7.1   Introduction

Mechanical parts come in many different physical shapes. The actual shape of any part effects how the stress develops within the part. Simple geometric forms like circles, squares, triangles and the like have been studied thoroughly and their properties are well-documented.

7.2   Problem

For the shapes to be studied, the following characteristics need to be discovered.
- Area
- Circumference
- Centroid
- Area moment of inertia

Function files have been written to find values for these shapes.
- Quarter, half, and full circles
- Rectangles
- Triangles

7.3   Theory

Each of the shapes presented have a few characteristic values, like width and radius, and one important point called the datum. The datum will be used when simple shapes are combined into composites. It is also the point from which the dimensions to the centroid are measured. Also note that moments of inertia are given about the centroidal axis.

Full Circles *circle (r, request)*: The only characteristic value of a circle is the radius. The datum of a circle is located in the center. The derived values of a circle are shown in Table 7.1.

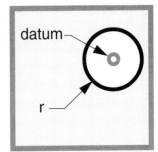

Characteristic	Value	Keyword
Area	πr^2	'area'
Circumference	$2\pi r$	'circ'
I_x	$\dfrac{\pi r^4}{4}$	'Ix'
I_y	$\dfrac{\pi r^4}{4}$	'Iy'
X distance to Centroid from datum	0	'centX'
Y distance to Centroid from datum	0	'centY'

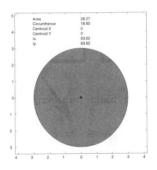

Table 7.1   Circle formulas.

Half Circles *halfcircle (r, orient, request)*:
    With half circles there are two characteristic values, the radius and the orientation. The datum is located at the center of the circle. There are four orientations of the half circle supported by the code. These orientations refer to which half of the circle is represented: north, east, south, or west. The derived values for half circles are given here for a north half

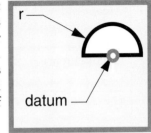

circle (Table 7.2). The values for the remaining semicircles are similar except the centroids may be located in the opposite direction. The possible values of the orientation are "n", "e", "s", and "w".

Characteristic	Value	Keyword
Area	$\dfrac{\pi r^2}{2}$	'area'
Circumference	$\pi r + 2r$	'circ'
I_x	$\dfrac{\pi r^4}{8}$	'Ix'
I_y	$\dfrac{\pi r^4}{8}$	'Iy'
X distance to Centroid from datum	0	'centX'
Y distance to Centroid from datum	$\dfrac{4r}{3\pi}$	'centY'

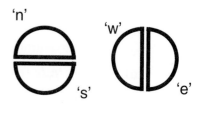

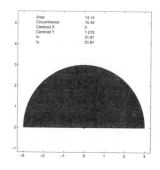

Table 7.2   Semicircle formulas for a north semicircle.

Quarter Circles *quartercircle (r, orient, request)*:

This last variation of the circle also has two characteristic values, the radius and the orientation. The datum is again located at what would be the center of the circle. The derived values for the northeast quarter circle are given in Table 7.3.

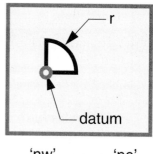

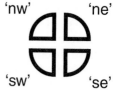

Characteristic	Value	Keyword
Area	$\dfrac{\pi r^2}{4}$	'area'
Circumference	$\dfrac{\pi r}{2} + 2r$	'circ'
I_x	$\dfrac{\pi r^4}{16}$	'Ix'

Table 7.3   Quarter circle formulas for a northeast quarter circle.

Characteristic	Value	Keyword
I_y	$\dfrac{\pi r^4}{16}$	'Iy'
X distance to Centroid from datum	$\dfrac{4r}{3\pi}$	'centX'
Y distance to Centroid from datum	$\dfrac{4r}{3\pi}$	'centY'

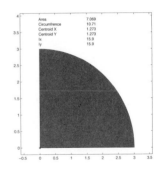

Table 7.3   Quarter circle formulas for a northeast quarter circle.

Rectangle *rectangle (b, h, request)*: This shape has two characteristic values, base and height. The datum is located in the lower left-hand corner of the shape. The calculated values for the rectangle are shown in Table 7.4.

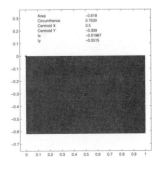

Characteristic	Value	Keyword
Area	bh	'area'
Circumference	$2(b+h)$	'circ'
I_x	$\dfrac{bh^3}{12}$	'Ix'
I_y	$\dfrac{hb^3}{12}$	'Iy'
X distance to Centroid from datum	$\dfrac{b}{2}$	'centX'
Y distance to Centroid from datum	$\dfrac{h}{2}$	'centY'

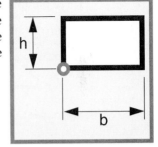

Table 7.4   Rectangle formulas.

Triangles *hortria (b, h, p, request) vertria (b, h, p, request)*: This shape is more problematic than the previous ones. So the triangle function has been broken into two different programs. The first, *hortria.m*, deals with triangles that have one edge that is perfectly horizontal. The second deals with triangles that have one edge that is perfectly vertical. Any triangle that has neither a horizontal or vertical edge can be broken down into two triangles that do, see Figure 7.1

hortria.m   *vertria.m*   2 x *vertria.m*

Figure 7.1   Different classifications of triangles.

Three parameters, see Figure 7.2, will be used to describe the triangle.

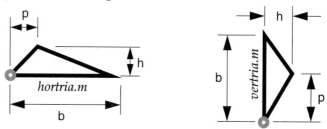

p and **h** may be negative. **b** must be positive.

Figure 7.2   Dimensioning of triangles.

The values of the dimensions follow the convention of positive values to the right or upward, negatives to the left or downward. Finding the characteristic values for triangles is difficult, even when the triangles have been broken into two categories, vertical and horizontal. The horizontal triangle is the only one that will be discussed, the vertical triangle follows the same methodology with some switching between the x and y axis. Table 7.5 lists the easy values for the horizontal triangle.

Characteristic	Value	Keyword
Area	$\dfrac{bh}{2}$	'area'

Table 7.5   Horizontal triangle formulas.

Characteristic	Value	Keyword
Circumference	$b + \sqrt{h^2 + p^2} + \sqrt{h^2 + (b - p)^2}$	'circ'
I_x	$\dfrac{bh^3}{36}$	'Ix'
I_y	Complicated	'Iy'
X distance to Centroid from datum	Complicated	'centX'
Y distance to Centroid from datum	$\dfrac{h}{3}$	'centY'

Table 7.5   Horizontal triangle formulas.

The difficulties arise in the triangle formulations because the point of the triangle can be located in four distinct regions, see Figure 7.3, all of which require a different way of solving. These values for the area moment of inertia and the distance to the centroid are not conveniently defined along both axes.

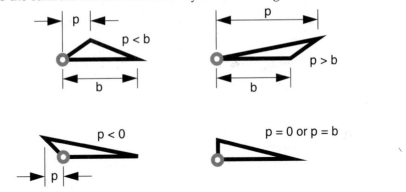

Figure 7.3   Distinct cases of a horizontal triangle.

The formulations for I_y and distance to the centroid along the x-axis differ depending on the case being dealt with. In the right triangle case, the formulas are simple, as shown in Table 7.6.

Characteristic	Value	Keyword
I_y	$\dfrac{hb^3}{36}$	'Iy'
X distance to Centroid from datum	$\dfrac{b}{3}$	'centX'

Table 7.6   Horizontal $p=0$ right triangle formulas.

The procedure for the remaining three cases is to break the original triangle into two new triangles sharing a common vertical edge. Using the dimensions from the original horizontal triangle, the dimensions of the two new triangles can be expressed, see Table 7.7.

Triangle type	Common base	Left triangle height	Left triangle point	Right triangle height	Right triangle point
	$\dfrac{bh}{\lvert p\rvert + b}$	p	h	b	0
	h	$-p$	0	$b-p$	0
	$\dfrac{bh}{p}$	$-b$	0	$p-b$	h

Table 7.7   Decomposed horizontal triangle values.

The horizontal centroids are easily calculated and combined using the two new vertical triangles. The area moments of inertia on the y axis are easily calculated. They can then be added and moved to the centroid using the parallel axis theorem.

This type of reasoning is applied to the vertical triangles also, and is equally valid there. Looking at the code it can be seen that each of the two triangle routines calls for information from the other. For instance, the "centX" section of the horizontal triangle routine calls for "centX" information from the vertical triangle routine. It seems like circular reference might occur, but it will not.

The actual application of all these equations is contained within the various routines. Like all the other routines in this book, it is not strictly necessary to understand all of the code to use them. The templates are printed here to give more information to those readers that are interested in coding, or for those that want to modify the routines to a more specific application.

7.4   Templates

circle.m

```
function [result]=circle(r,req)
%CIRCLE Circle shape routine.
%    CIRCLE(RADIUS, REQUEST)
%
%    Shape described: Circle.
%
%    Datum location: Center.
%
%    Input arguments:
%       radius: Radius of circle.
%
%    Requests: (Must be in single quotes)
%       'area':  Area of the shape.
%       'circ':  Circumference of shape.
%       'Ix':    Area moment of inertia about the neutral x axis.
%       'Iy':    Area moment of inertia about the neutral y axis.
%       'centX': Distance from datum to centroid in the x direction.
%       'centY': Distance from datum to centroid in the y direction.
%       'J':     Polar moment of inertia.
%       'comp':  All of the above in a 1x6 matrix.
%       'draw':  Show the shape graphically.
%
%    See also CIRCLE, COMP, HALFCIRCLE, HORTRAP, HORTRIA, IBEAM,
LBEAM,
%        OBEAM, QUARTERCIRCLE, RECTANGLE, RECTUBE, TBEAM, VERTRAP,
VERTRIA.

%    Details are to be found in Mastering Mechanics I, Douglas W.
Hull,
%    Prentice Hall, 1998

%    Douglas W. Hull, 1998
%    Copyright (c) 1998-99 by Prentice Hall
%    Version 1.00

req=lower(req);
```

```
if      strcmp(req,'area')   result=pi*r.^2;
elseif strcmp(req,'circ')   result=2*pi*r;
elseif strcmp(req,'ix')     result=r.^4*pi/4;
elseif strcmp(req,'iy')     result=r.^4*pi/4;
elseif strcmp(req,'centx')  result=0;
elseif strcmp(req,'centy')  result=0;
elseif strcmp(req,'j')      result=pi/2*r.^4;
elseif strcmp(req,'comp')
  result(1,1)=circle(r,'area');
  result(1,2)=circle(r,'circ');
  result(1,3)=circle(r,'centx');
  result(1,4)=circle(r,'centx');
  result(1,5)=circle(r,'ix');
  result(1,6)=circle(r,'iy');
elseif strcmp(req,'draw')
  xcentroid=circle(r,'centx');
  ycentroid=circle(r,'centy');
  angle =[0: DR(1): DR(360)];
  X=r*cos(angle);
  Y=r*sin(angle);
  fill(X,Y,'r')
  hold on
  ColA=strvcat('Area','Circumference','Centroid X','Centroid
Y','Ix','Iy');
  ColB=makecol(circle(r,'comp')');
  plot (xcentroid,ycentroid,'ko',0,0,'g*')
  hold off;
  axis ('equal')
  titleblock(ColA,ColB)
  expandaxis(5,5,5,0)
else
  disp ('That is not a valid request, try area, circ, Ix, Iy,
centX,')
  disp ('centY, comp.')
  disp ('You must use single quotes')
end
```

halfcircle.m

```
function [result]=halfcircle(r,orient,req)
%HALFCIRCLE Semicircle shape routine.
%    HALFCIRCLE(RADIUS,ORIENT,REQUEST)
%
%    Shape described: A semi-circle.
%
%    Datum location: Center of circle.
%
%    Input arguments:
%      RADIUS: Radius of circle.
%      ORIENTATION: Either 'n','e','s', or 'w'.
%
%    Requests: (Must be in single quotes)
%      'area':  Area of the shape.
%      'circ':  Circumference of shape.
%      'Ix':    Area moment of inertia about the neutral x axis.
%      'Iy':    Area moment of inertia about the neutral y axis.
%      'centX': Distance from datum to centroid in the x direction.
%      'centY': Distance from datum to centroid in the y direction.
%      'comp':  All of the above in a 1x6 matrix.
%
%    See also CHANNEL, CIRCLE, COMP, HORTRAP, HORTRIA, IBEAM, LBEAM,
%      OBEAM, QUARTERCIRCLE, RECTANGLE, RECTUBE, TBEAM, VERTRAP, VERTRIA.

%    Details are to be found in Mastering Mechanics I, Douglas W. Hull,
%    Prentice Hall, 1999

%    Douglas W. Hull, 1999
%    Copyright (c) 1999 by Prentice Hall
%    Version 1.00

req=lower(req);
orient=lower(orient);

if     strcmp(req,'area')  result=(pi*r^2)/2;
elseif strcmp(req,'circ')  result=(2*pi*r)/2+2*r;
elseif strcmp(req,'ix')    result=r^4*pi/8;
elseif strcmp(req,'iy')    result=r^4*pi/8;
elseif strcmp(req,'centx')
  if strcmp(orient,'w') result=-4*r/(3*pi);
  elseif strcmp(orient,'n') | strcmp(orient,'s')
    result=0;
  else
    result=4*r/(3*pi);
  end
elseif strcmp(req,'centy')
  if strcmp(orient,'s') result=-4*r/(3*pi);
  elseif strcmp(orient,'e') | strcmp(orient,'w')
```

```
      result=0;
   else
      result=4*r/(3*pi);
   end
elseif strcmp(req,'comp')
   result(1,1)=halfcircle(r,orient,'area');
   result(1,2)=halfcircle(r,orient,'circ');
   result(1,3)=halfcircle(r,orient,'centx');
   result(1,4)=halfcircle(r,orient,'centy');
   result(1,5)=halfcircle(r,orient,'ix');
   result(1,6)=halfcircle(r,orient,'iy');
elseif strcmp(req,'draw')
   if strcmp(orient,'n')
      angle =[0: DR(1): DR(180)];
   elseif strcmp(orient,'e')
      angle =[DR(-90): DR(1): DR(90)];
   elseif strcmp(orient,'s')
      angle =[DR(180): DR(1): DR(360)];
   else strcmp(orient,'w')
      angle =[DR(90): DR(1): DR(270)];
   end
   xcentroid=halfcircle(r,orient,'centx');
   ycentroid=halfcircle(r,orient,'centy');
   X=r*cos(angle);
   Y=r*sin(angle);
   fill(X,Y,'r');
   hold on
   ColA=strvcat('Area','Circumference','Centroid X','Centroid
Y','Ix','Iy');
   ColB=makecol(halfcircle(r,orient,'comp')');
   plot (xcentroid,ycentroid,'ko',0,0,'g*')
   hold off;
   axis ('equal')
   titleblock(ColA,ColB)
   expandaxis(5,5,5,0)
else
   disp ('That is not a valid request, try area, circ, Ix, Iy,
centX,')
   disp ('centY, comp.')
   disp ('You must use single quotes')
end
```

quartercircle.m

```
function [result]=quartercirc(r,orient,req)
%QUARTERCIRCLE Quarter circle shape routine.
%    QUARTERCIRCLE(radius,request)
%
%    Shape described: Quarter circle.
%
%    Datum location: Center of circle.
%
%    Input arguments:
%      RADIUS: Radius of circle.
%
%    Requests: (Must be in single quotes)
%      'area':  Area of the shape.
%      'circ':  Circumference of shape.
%      'Ix':    Area moment of inertia about the neutral x axis.
%      'Iy':    Area moment of inertia about the neutral y axis.
%      'centX': Distance from datum to centroid in the x direction.
%      'centY': Distance from datum to centroid in the y direction.
%      'comp':  All of the above in a 1x6 matrix.
%      'draw':  Show the shape graphically.
%
%    See also CHANNEL, CIRCLE, COMP, HALFCIRCLE, HORTRAP, HORTRIA, IBEAM,
%       LBEAM, OBEAM, RECTANGLE, RECTUBE, TBEAM, VERTRAP, VERTRIA.

%    Details are to be found in Mastering Mechanics I, Douglas W. Hull,
%    Prentice Hall, 1999

%    Douglas W. Hull, 1999
%    Copyright (c) 1999 by Prentice Hall
%    Version 1.00

req=lower(req);
orient=lower(orient);

if     strcmp(req,'area')   result=(pi*r^2)/4;
elseif strcmp(req,'circ')   result=(2*pi*r)/4+2*r;
elseif strcmp(req,'ix')     result=r^4*pi/16;
elseif strcmp(req,'iy')     result=r^4*pi/16;
elseif strcmp(req,'centx')
  if strcmp(orient,'nw') | strcmp(orient,'sw')
     result=-4*r/(3*pi);
  else
     result=4*r/(3*pi);
  end
elseif strcmp(req,'centy')
  if strcmp(orient,'se') | strcmp(orient,'sw')
     result=-4*r/(3*pi);
  else
```

```
      result=4*r/(3*pi);
   end
elseif strcmp(req,'comp')
   result(1,1)=quartercircle(r,orient,'area');
   result(1,2)=quartercircle(r,orient,'circ');
   result(1,3)=quartercircle(r,orient,'centx');
   result(1,4)=quartercircle(r,orient,'centx');
   result(1,5)=quartercircle(r,orient,'ix');
   result(1,6)=quartercircle(r,orient,'iy');
elseif strcmp(req,'draw')
   if strcmp(orient,'ne')
     angle =[0: DR(1): DR(90)];
   elseif strcmp(orient,'se')
     angle =[DR(-90): DR(1): DR(0)];
   elseif strcmp(orient,'sw')
     angle =[DR(180): DR(1): DR(270)];
   else strcmp(orient,'nw')
     angle =[DR(90): DR(1): DR(180)];
   end
   xcentroid=quartercircle(r,orient,'centx');
   ycentroid=quartercircle(r,orient,'centy');
   X=[0, r*cos(angle), 0];
   Y=[0, r*sin(angle), 0];
   fill(X,Y,'r')
   hold on
   ColA=strvcat('Area','Circumference','Centroid X','Centroid
Y','Ix','Iy');
   ColB=makecol(quartercircle(r,orient,'comp')');
   plot (xcentroid,ycentroid,'ko',0,0,'g*')
   hold off;
   axis ('equal')
   titleblock(ColA,ColB)
   expandaxis(5,5,5,0)
else
   disp ('That is not a valid request, try area, circ, Ix, Iy,
centX,')
   disp ('centY, comp.')
   disp ('You must use single quotes')
end
```

rectangle.m

```
function [result]=rectangle(b,h,req)
%RECTANGLE Rectangular shape routine.
%    [result]=VERTRAP(BASE,HEIGHT,REQUEST)
%
%    Shape described: A rectangle.
%
%    Datum location: Bottom left corner.
%
%    Input arguments:
%      base: Distance between the vertical sides.
%      height: Distance between the horizontal sides.
%
%    Requests: (Must be in single quotes)
%      'area':  Area of the shape.
%      'circ':  Circumference of shape.
%      'Ix':    Area moment of inertia about the neutral x axis.
%      'Iy':    Area moment of inertia about the neutral y axis.
%      'centX': Distance from datum to centroid in the x direction.
%      'centY': Distance from datum to centroid in the y direction.
%      'comp':  All of the above in a 1x6 matrix.
%      'draw':  Show the shape graphically.
%
%    See also CHANNEL, CIRCLE, COMP, HALFCIRCLE, HORTRAP, HORTRIA, IBEAM,
%       LBEAM, OBEAM, QUARTERCIRCLE, RECTUBE, TBEAM, VERTRAP, VERTRIA.

%    Details are to be found in Mastering Mechanics I, Douglas W. Hull,
%    Prentice Hall, 1999

%    Douglas W. Hull, 1999
%    Copyright (c) 1999 by Prentice Hall
%    Version 1.00

req=lower(req);

if     strcmp(req,'area')  result=b*h;
elseif strcmp(req,'circ')  result=2*(b+h);
elseif strcmp(req,'ix')    result=b*h^3/12;
elseif strcmp(req,'iy')    result=h*b^3/12;
elseif strcmp(req,'centx') result=b/2;
elseif strcmp(req,'centy') result=h/2;
elseif strcmp(req,'comp')
  result(1,1)=rectangle(b,h,'area');
  result(1,2)=rectangle(b,h,'circ');
  result(1,3)=rectangle(b,h,'centx');
  result(1,4)=rectangle(b,h,'centy');
  result(1,5)=rectangle(b,h,'ix');
  result(1,6)=rectangle(b,h,'iy');
elseif strcmp(req,'draw')
```

```
   xcentroid=rectangle(b,h,'centx');
   ycentroid=rectangle(b,h,'centy');
   X=[0 b b 0 0];
   Y=[0 0 h h 0];
   fill(X,Y,'r')
   hold on;
   ColA=strvcat('Area','Circumference','Centroid X','Centroid
Y','Ix','Iy');
   ColB=makecol(rectangle(b,h,'comp')');
   plot (xcentroid,ycentroid,'ko',0,0,'g*')
   hold off
   axis ('equal')
   titleblock(ColA,ColB)
   expandaxis(5,5,5,0)
else
   disp ('That is not a valid request, try area, circ, Ix, Iy,
centX,')
   disp ('centY, comp.')
   disp ('You must use single quotes')
end
```

hortria.m

```
function [result]=hortria(b,h,p,req)
%HORTRIA Horizontal triangle shape routine.
%    HORTRIA(BASE,HEIGHT,P,REQUEST)
%
%    Shape described: Triangle with a horizontal base.
%
%    Datum location: Leftmost point of base.
%
%    Input arguments:
%      BASE: Length of base (base must be a horizontal line).
%      HEIGHT: Vertical distance between base and vertex may be negative.
%      P: Horizontal distance from the datum to lower edge of small
edge. May
%        be negative.
%
%    Requests: (Must be in single quotes)
%      'area':  Area of the shape.
%      'circ':  Circumference of shape.
%      'Ix':    Area moment of inertia about the neutral x axis.
%      'Iy':    Area moment of inertia about the neutral y axis.
%      'centX': Distance from datum to centroid in the x direction.
%      'centY': Distance from datum to centroid in the y direction.
%      'comp':  All of the above in a 1x6 matrix.
%      'draw':  Show the shape graphically.
```

```
%
%   See also CHANNEL, CIRCLE, COMP, HALFCIRCLE, HORTRAP, IBEAM, LBEAM,
%     OBEAM, QUARTERCIRCLE, RECTANGLE, RECTUBE, TBEAM, VERTRAP, VERTRIA.

%   Details are to be found in Mastering Mechanics I, Douglas W. Hull,
%   Prentice Hall, 1999

%   Douglas W. Hull, 1999
%   Copyright (c) 1999 by Prentice Hall
%   Version 1.00

req=lower(req);

origH=h;
h=abs(h);
if b<0
  disp ('Base must be a positive length')
  return
end

if      strcmp(req,'area')  result=b*h/2;
elseif strcmp(req,'circ')  result=sqrt(h^2+p^2)+sqrt(h^2+(b-p)^2)+b;
elseif strcmp(req,'ix')     result=b*h^3/36;
elseif strcmp(req,'iy')
  if p==0 | p==b % right triangle
    result=h*b^3/36;
  end
  if p<0  % vertex extend past left edge of base.
    midh=b*h/(abs(p)+b);
    lsi=vertria(midh,p,h,'iy');
    lsa=vertria(midh,p,h,'area');
    lsd=vertria(midh,p,h,'centX');
    rsi=vertria(midh,b,0,'iy');
    rsa=vertria(midh,b,0,'area');
    rsd=vertria(midh,b,0,'centx');
    centroid=hortria(b,h,p,'centx');
    result=lsi+lsa*(centroid-lsd)^2+rsi+rsa*(centroid-rsd)^2;
  end
  if p>b  % vertex extend past right edge of base.
    midh=b*h/p;
    lsi=vertria(midh,-b,0,'iy');
    lsa=vertria(midh,-b,0,'area');
    lsd=vertria(midh,-b,0,'centx')+b;
    rsi=vertria(midh,p-b,h,'iy');
    rsa=vertria(midh,p-b,h,'area');
    rsd=vertria(midh,p-b,h,'centx')+b;
    centroid=hortria(b,h,p,'centx');
    result=lsi+lsa*(centroid-lsd)^2+rsi+rsa*(centroid-rsd)^2;
  end
```

```
    if p>0 & p<b % vertex is over the base
      lsi=vertria(h,-p,0,'iy');
      lsa=vertria(h,-p,0,'area');
      lsd=vertria(h,-p,0,'centx')+p;
      rsi=vertria(h,b-p,0,'iy');
      rsa=vertria(h,b-p,0,'area');
      rsd=vertria(h,b-p,0,'centx')+p;
      centroid=hortria(b,h,p,'centx');
      result=lsi+lsa*(centroid-lsd)^2+rsi+rsa*(centroid-rsd)^2;
    end

elseif strcmp(req,'centx')
  if p<0  % vertex extend past left edge of base.
    midh=b*h/(abs(p)+b);
    lsa=vertria(midh,p,h,'area');
    lsd=vertria(midh,p,h,'centx');
    rsa=vertria(midh,b,0,'area');
    rsd=vertria(midh,b,0,'centx');
    wa=hortria(b,h,p,'area');
    result=((lsa*lsd)+(rsa*rsd))/wa;
  end
if p==0 % right triangle
  result=b/3;
end
if p==b % right triangle
  result=b*2/3;
end
if p>b  % vertex extend past right edge of base.
  midh=b*h/p;
  lsa=vertria(midh,-b,0,'area');
  lsd=vertria(midh,-b,0,'centx')+b;
  rsa=vertria(midh,p-b,h,'area');
  rsd=vertria(midh,p-b,h,'centx')+b;
  wa=hortria(b,h,p,'area');
  result=((lsa*lsd)+(rsa*rsd))/wa;
end
if p>0 & p<b % vertex is over the base
  lsa=vertria(h,-p,0,'area');
  lsd=vertria(h,-p,0,'centx')+p;
  rsa=vertria(h,b-p,0,'area');
  rsd=vertria(h,b-p,0,'centx')+p;
  wa=hortria(b,h,p,'area');
  result=((lsa*lsd)+(rsa*rsd))/wa;
end
elseif strcmp(req,'centy')
  result=origH/3;
elseif strcmp(req,'comp')
```

```
    result(1,1)=hortria(b,origH,p,'area');
    result(1,2)=hortria(b,origH,p,'circ');
    result(1,3)=hortria(b,origH,p,'centx');
    result(1,4)=hortria(b,origH,p,'centy');
    result(1,5)=hortria(b,origH,p,'ix');
    result(1,6)=hortria(b,origH,p,'iy');
elseif strcmp(req,'draw')
   xcentroid=hortria(b,origH,p,'centx');
   ycentroid=hortria(b,origH,p,'centy');
   X=[0 b p 0];
   Y=[0 0 origH 0];
   fill(X,Y,'r')
   hold on;
   ColA=strvcat('Area','Circumference','Centroid X','Centroid
Y','Ix','Iy');
   ColB=makecol(hortria(b,origH,p,'comp')');
   plot (xcentroid,ycentroid,'ko',0,0,'g*')
   hold off
   axis ('equal')
   titleblock(ColA,ColB)
   expandaxis(5,5,5,0)
else
   disp ('That is not a valid request, try area, circ, Ix, Iy,
centX,')
   disp ('centY, comp.')
   disp ('You must use single quotes')
end
```

---

vertria.m

```
function [result]=vertria(b,h,p,req)
%VERTRIA Horizontal triangle shape routine.
%    VERTRIA(BASE,HEIGHT,P,REQUEST)
%
%    Shape described: Triangle with a verizontal base.
%
%    Datum location: Leftmost point of base.
%
%    Input arguments:
%      BASE: Length of base (base must be a horizontal line).
%      HEIGHT: Vertical distance between base and vertex may be negative.
%      P: Horizontal distance from the datum to lower edge of small
edge. May
%        be negative.
%
```

```
%    Requests: (Must be in single quotes)
%      'area':  Area of the shape.
%      'circ':  Circumference of shape.
%      'Ix':    Area moment of inertia about the neutral x axis.
%      'Iy':    Area moment of inertia about the neutral y axis.
%      'centX': Distance from datum to centroid in the x direction.
%      'centY': Distance from datum to centroid in the y direction.
%      'comp':  All of the above in a 1x6 matrix.
%      'draw':  Show the shape graphically.
%
%    See also CHANNEL, CIRCLE, COMP, HALFCIRCLE, HORTRAP, IBEAM, LBEAM,
%        OBEAM, QUARTERCIRCLE, RECTANGLE, RECTUBE, TBEAM, VERTRAP.


%    Details are to be found in Mastering Mechanics I, Douglas W. Hull,
%    Prentice Hall, 1999


%    Douglas W. Hull, 1999
%    Copyright (c) 1999 by Prentice Hall
%    Version 1.00

req=lower(req);

origH=h;
h=abs(h);
if b<0
  disp ('Base must be a positive length')
  return
end

if      strcmp(req,'area')   result=b*h/2;
elseif strcmp(req,'circ')   result=sqrt(h^2+p^2)+sqrt(h^2+(b-
p)^2)+b;
elseif strcmp(req,'iy')     result=b*h^3/36;
elseif strcmp(req,'ix')
  if p==0 | p==b % right triangle
    result=h*b^3/36;
  end
  if p<0 % vertex extends below base
    midh=b*h/(abs(p)+b);
    lsi=hortria(midh,p,h,'ix');
    lsa=hortria(midh,p,h,'area');
    lsd=hortria(midh,p,h,'centy');
    usi=hortria(midh,b,0,'ix');
    usa=hortria(midh,b,0,'area');
    usd=hortria(midh,b,0,'centy');
    centroid=vertria(b,h,p,'centy');
    result=lsi+lsa*(centroid-lsd)^2+usi+usa*(centroid-usd)^2;
  end
```

```
    if p>b % vertex extends above base
      midh=b*h/p;
      lsi=hortria(midh,-b,0,'ix');
      lsa=hortria(midh,-b,0,'area');
      lsd=hortria(midh,-b,0,'centy')+b;
      usi=hortria(midh,p-b,h,'ix');
      usa=hortria(midh,p-b,h,'area');
      usd=hortria(midh,p-b,h,'centy')+b;
      centroid=vertria(b,h,p,'centy');
      result=lsi+lsa*(centroid-lsd)^2+usi+usa*(centroid-usd)^2;
    end
    if p>0 & p<b % vertex is across from the base
      lsi=hortria(h,-p,0,'ix');
      lsa=hortria(h,-p,0,'area');
      lsd=hortria(h,-p,0,'centy')+p;
      usi=hortria(h,b-p,0,'ix');
      usa=hortria(h,b-p,0,'area');
      usd=hortria(h,b-p,0,'centy')+p;
      centroid=vertria(b,h,p,'centy');
      result=lsi+lsa*(centroid-lsd)^2+usi+usa*(centroid-usd)^2;
    end

elseif strcmp(req,'centy')
    if p<0 % vertex extends below base
      midh=b*h/(abs(p)+b);
      lsa=hortria(midh,p,h,'area');
      lsd=hortria(midh,p,h,'centy');
      usa=hortria(midh,b,0,'area');
      usd=hortria(midh,b,0,'centy');
      wa=vertria(b,h,p,'area');
      result=((lsa*lsd)+(usa*usd))/wa;
    end
    if p==0 % right triangle
      result=b/3;
    end
    if p==b % right triangle
      result=b*2/3;
    end
    if p>b % vertex extends above base
      midh=b*h/p;
      lsa=hortria(midh,-b,0,'area');
      lsd=hortria(midh,-b,0,'centy')+b;
      usa=hortria(midh,p-b,h,'area');
      usd=hortria(midh,p-b,h,'centy')+b;
      wa=vertria(b,h,p,'area');
      result=((lsa*lsd)+(usa*usd))/wa;
    end
    if p>0 & p<b % vertex is across from the base
```

```
      lsa=hortria(h,-p,0,'area');
      lsd=hortria(h,-p,0,'centy')+p;
      usa=hortria(h,b-p,0,'area');
      usd=hortria(h,b-p,0,'centy')+p;
      wa=vertria(b,h,p,'area');
      result=((lsa*lsd)+(usa*usd))/wa;
   end
elseif strcmp(req,'centx')
   result=origH/3;
elseif strcmp(req,'comp')
   result(1,1)=vertria(b,origH,p,'area');
   result(1,2)=vertria(b,origH,p,'circ');
   result(1,3)=vertria(b,origH,p,'centx');
   result(1,4)=vertria(b,origH,p,'centy');
   result(1,5)=vertria(b,origH,p,'ix');
   result(1,6)=vertria(b,origH,p,'iy');
elseif strcmp(req,'draw')
   xcentroid=vertria(b,origH,p,'centx');
   ycentroid=vertria(b,origH,p,'centy');
   X=[0 origH 0 0];
   Y=[0 p b 0];
   fill(X,Y,'r')
   hold on;
   ColA=strvcat('Area','Circumference','Centroid X','Centroid
Y','Ix','Iy');
   ColB=makecol(vertria(b,origH,p,'comp')');
   plot (xcentroid,ycentroid,'ko',0,0,'g*')
   hold off
   axis ('equal')
   titleblock(ColA,ColB)
   expandaxis(5,5,5,0)
else
   disp ('That is not a valid request, try area, circ, Ix, Iy,
centX,')
   disp ('centY, comp.')
   disp ('You must use single quotes')
end
```

7.5   Output

These routines can give much of the engineering information needed from simple shapes. The following examples will show how to describe many different shapes and how to find information about them.

For each of the following shapes, find the
- Area
- Circumference
- Centroid
- I_x and I_y

```
>>Aarea=rectangle(8,2,'area')
Aarea =
        16
>>Acirc=rectangle(8,2,'circ')
Acirc =
        20
>>Acentx=rectangle(8,2,'centX')
Acentx =
        4
>>Acenty=rectangle(8,2,'centY')
Acenty =
        1
>>AIx=rectangle(8,2,'Ix')
AIx =
    5.3333
>>AIy=rectangle(8,2,'Iy')
AIy =
    85.3333
>>rectangle(8,2,'draw')
```

a.)

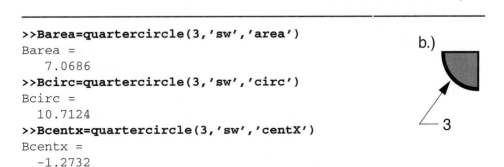

```
>>Barea=quartercircle(3,'sw','area')
Barea =
    7.0686
>>Bcirc=quartercircle(3,'sw','circ')
Bcirc =
    10.7124
>>Bcentx=quartercircle(3,'sw','centX')
Bcentx =
    -1.2732
```

b.)

```
>>Bcenty=quartercircle(3,'sw','centY
')
Bcenty =
  -1.2732
>>BIx=quartercircle(3,'sw','Ix')
BIx =
  15.9043
>>BIy=quartercircle(3,'sw','Iy')
BIy =
  15.9043
>>quartercircle(3,'sw','draw')
```

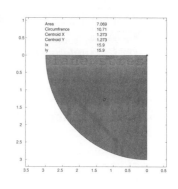

```
>>Carea=hortria(8,-3,-4,'area')
Carea =
       12
>>Ccirc=hortria(8,-3,-4,'circ')
Ccirc =
  25.3693
>>Ccentx=hortria(8,-3,-
4,'centX')
Ccentx =
   1.3333
>>Ccenty=hortria(8,-3,-4,'centY')
Ccenty =
       -1
>>CIx=hortria(8,-3,-4,'Ix')
CIx =
        6
>>CIy=hortria(8,-3,-4,'Iy')
CIy =
  74.6667
>>hortria(8,-3,-4,'draw')
```

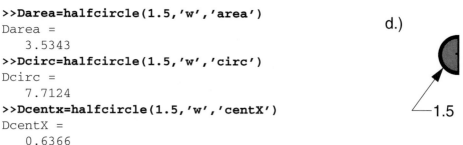

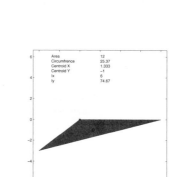

```
>>Darea=halfcircle(1.5,'w','area')
Darea =
   3.5343
>>Dcirc=halfcircle(1.5,'w','circ')
Dcirc =
   7.7124
>>Dcentx=halfcircle(1.5,'w','centX')
DcentX =
   0.6366
```

d.)

1.5

```
>>Dcenty=halfcircle(1.5,'w','centY')
DcentY =
         0
>>DIx=halfcircle(1.5,'w','Ix')
DIx =
    1.9880
>>DIy=halfcircle(1.5,'w','Iy')
DIy =
    1.9880
>>halfcircle(1.5,'w','draw')
```

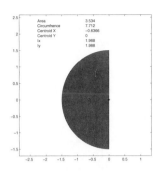

```
>>Earea=vertria(5,-3,2,'area')
Earea =
    7.5000
>>Ecirc=vertria(5,-3,2,'circ')
Ecirc =
    12.8482
>>Ecentx=vertria(5,-3,2,'centX')
Ecentx =
        -1
>>Ecenty=vertria(5,-3,2,'centY')
Ecenty =
    2.3333
>>EIx=vertria(5,-3,2,'Ix')
EIx =
    7.9167
>>EIy=vertria(5,-3,2,'Iy')
EIy =
    3.7500
>>vertria(5,-3,2,'draw')
```

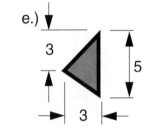

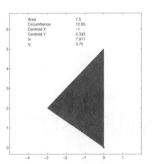

```
>>Farea=circle(4,'area')
Farea =
    50.2655
>>Fcirc=circle(4,'circ')
Fcirc =
    25.1327
>>Fcentx=circle(4,'centX')
Fcentx =
         0
>>Fcenty=circle(4,'centY')
```

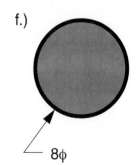

```
Fcenty =
          0
>>FIx=circle(4,'Ix')
FIx =
  201.0619
>>FIy=circle(4,'Iy')
FIy =
  201.0619
>>circle(4,'draw')
```

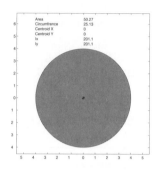

These are but six simple examples of the possibilities for these shape routines.

7.6   Features

Some good features about these routines are their flexibility in the amount of different information they can return, all the important engineering values can be retrieved with one function.

Readers who study the code will notice a feature of these functions that has not yet been explored, the "comp" call. This keyword is short for composition and will be covered in detail in Chapter 8.

7.7   Summary

Required argument

[answer]=circle (*Radius,'keyword'*)

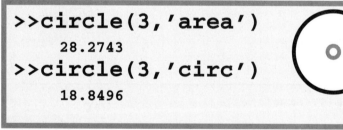

[answer]=halfcirc (*Radius,orientation,'keyword'*)

>>halfcircle(3,'n','area')
   14.1372
>>halfcircle(3,'n','centY')
   1.2732

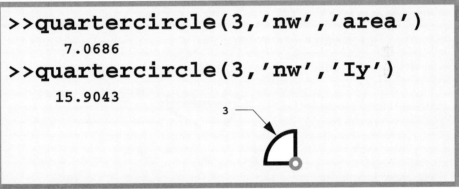

[answer]=nwcircle (*Radius,'keyword'*)[answer]=rectangl

>>quartercircle(3,'nw','area')
   7.0686
>>quartercircle(3,'nw','Iy')
   15.9043

(*Base,Height,'keyword'*)

>>rectangle(6,4,'area')
   24
>>rectangle(6,4,'Iy')
   72

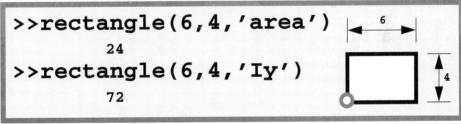

[answer]=hortria (*Base,Height,Point,'keyword'*)

>>**hortria(7,3,2,'area')**
   20.5000
>>**hortria(7,3,2,'Iy')**
   22.7500

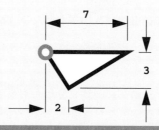

>>**hortria(7,-3,2,'area')**
   10.5000
>>**hortria(7,-3,2,'Iy')**
   22.7500

>>**hortria(7,3,-2,'area')**
   10.5000
>>**hortria(7,3,-2,'Iy')**
   39.0833

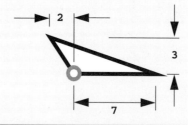

[answer]=vertria (*Base,Height,Point,'keyword'*)

```
>>hortria(7,3,0,'area')
     10.5000
>>hortria(7,3,0,'Iy')
     28.5833
>>vertria(3,7,0,'area')
     10.5000
>>vertria(3,7,0,'Iy')
     28.5833
```

```
>>vertria(7,3,2,'area')
     10.5000
>>vertria(7,3,2,'Iy')
      5.2500
```

```
>>vertria(7,-3,2,'area')
     10.5000
>>vertria(7,-3,2,'Iy')
      5.2500
```

```
>>vertria(7,-3,-2,'area')
     10.5000
>>vertria(7,-3,-2,'Iy')
      5.2500
```

The "keyword" argument for these functions must include the single quotes and can be chosen from this list:

- 'area' the area of the shape.
- 'circ' the length of the perimeter of the shape.
- 'centX' the distance from the datum to the centroid along the x axis.
- 'centY' the distance from the datum to the centroid along the y axis.
- 'Ix' the area moment of inertia about the x axis through the centroid.
- 'Iy' the area moment of inertia about the y axis through the centroid.
- 'comp' all of the above into a 1x6 matrix. This is useful for composite shapes that are covered in the next chapter.
- 'draw' the graphical representation of the shape.

8

Geometry of Standard Composite Shapes

8.1   Introduction

The primitive shapes explored in the previous chapter are mathmatically well defined and are often sufficient models of the real shape that engineers must deal with. When the simple shapes are not sufficient, these well-known geometric forms can be combined into new forms that can also be studied. These composite shapes are often close enough to the true physical parts that model and corresponding results are valid.

8.2   Problem

For the shapes to be studied, the following characteristics need to be discovered.
- Area
- Circumference
- Area moment of inertia
- Centroid

Function files have been written to find values for these shapes.
- Angle
- Round tube
- Rectangular tube
- Channel
- I-beams
- T-beams
- Trapezoids

8.3   Theory

The primitive shapes of the previous chapter had few characteristic dimensions. Likewise these composite shapes have more characteristic values, because of the added complexity. These shapes all have a datum that the measurements are made from.

Angle *lbeam (hl, vl, ht, vt, orient, request)*:

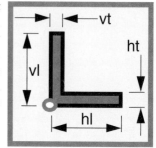

The angle cross-section has four parameters that describe it's shape, and one that describes orientation. The orientation parameter can be remembered by thinking about what direction the angle is pointing at. The example here is pointing to the southwest, so the parameter would be "sw." Because the function *angle.m* is already defined in MATLAB, the name *lbeam.m* was chosen.

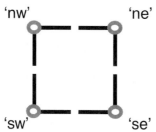

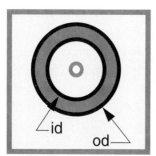

Round Tube *obeam (od, id, request)*: This tube cross-section has only two parameters so it is pretty simple. However, unlike the angle cross-section it has both a positive element, and a negative one. The negative cross-section represents the hole. This is all taken care of by the routine, so there is no need to worry about the specifics. Again, the preferred name for this function, *tube.m* was already defined in MATLAB so the name *obeam.m* was chosen.

Rectangular Tube

rectube (ob, oh, ib, ih, request): This cross-section is similar to the round tube in that it also employs a negative component to represent the hole.

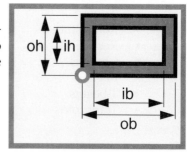

Channel *channel (b, h, bt, lt, request)*:

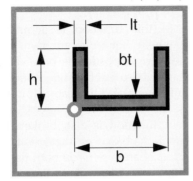

This cross-section is common in engineering applications. Along with the dimensions, there must also be an orientation parameter. The parameter is the direction that has the open end. This drawing shows an orientation of "n", north.

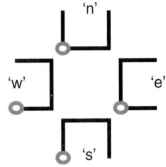

I-Beams *ibeam (b, h, bt, wt, orient, request)*:

This common cross-section employs not only the dimensions, but also an orientation parameter. The parameter is either "I" or "H" depending on which letter it resembles.

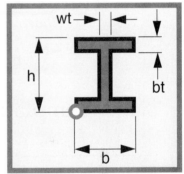

T-Beams *ibeam (b, h, bt, wt, orient, request)*:

This common cross-section employs not only the dimensions, but also an orientation parameter. The parameter is either 'I' or 'H' depending on which letter it resembles.

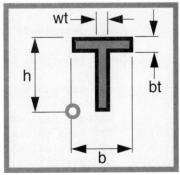

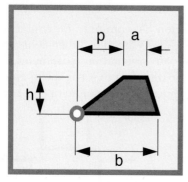

Trapezoids *hortrap (b, h, a, p, request)* *vertrap (b, h, a, p, request)*:
The trapezoid of this function is actually a simplified one where not only are two sides parallel, but the edge opposite the base cannot extend beyond the base itself. This is not an outrageous requirement, and it greatly simplifies the calculations.

This trapezoid function is also broken into two different functions based on the orientation of the base. *Hortrap.m* and *vertrap.m* both use the same characteristic dimensions but with a different orientation. The values *a, b, p* need to be positive, but the *h* value can be negative.

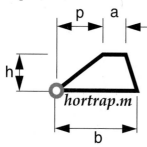
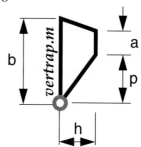

8.4   Template

lbeam.m

```
function [result]=lbeam(hl,vl,ht,vt,orient,req)
%LBEAM L-beam shape routine.
%    LBEAM(BASE,HEIGHT,BT,VT,ORIENT,REQUEST)
%
%    Shape described: L-beam.
%
%    Datum location: Outer corner.
%
%    Input arguments:
%      BASE: Distance between the vertical sides.
%      HEIGHT: Distance between the horizontal sides.
%      BT: Base thickness.
%      VT: Vertical thickness.
%      ORIENT: Either 'ne','se','sw', or 'nw'.
%
```

```
%    Requests: (Must be in single quotes)
%      'area':  Area of the shape.
%      'circ': Circumference of shape.
%      'Ix':    Area moment of inertia about the neutral x axis.
%      'Iy':    Area moment of inertia about the neutral y axis.
%      'centX': Distance from datum to centroid in the x direction.
%      'centY': Distance from datum to centroid in the y direction.
%      'comp':  All of the above in a 1x6 matrix.
%      'draw':  Show the shape graphically.
%
%    See also CHANNEL, CIRCLE, COMP, HALFCIRCLE, HORTRAP, HORTRIA, IBEAM,
%        OBEAM, QUARTERCIRCLE, RECTANGLE, RECTUBE, TBEAM, VERTRAP, VERTRIA.

%    Details are to be found in Mastering Mechanics I, Douglas W. Hull,
%    Prentice Hall, 1999

%    Douglas W. Hull, 1999
%    Copyright (c) 1999 by Prentice Hall
%    Version 1.00

req=lower(req);
orient=lower(orient);

if (ht < 0) | (vt < 0) | (hl < 0) | (vl < 0)
  disp ('None of the arguments may be negative')
  return
end
if (ht > vl) | (vt > hl)
  disp ('The thickness of one leg cannot be greater than the
length')
  disp ('of the opposite leg')
  return
end

% breaks up the angle into two rectangles to be used in the
% comp shape routine.
part(1,:)=[rectangle(hl,ht,'comp'),0,0,1];
part(2,:)=[rectangle(vt,(vl-ht),'comp'),0,ht,1];
if     strcmp(req,'area')  result=comp(part,'area');
elseif strcmp(req,'circ')  result=2*(hl+vl);
elseif strcmp(req,'Ix')  result=comp(part,'ix');
elseif strcmp(req,'Iy')  result=comp(part,'iy');
elseif strcmp(req,'centx')
  if strcmp(orient,'ne') | strcmp(orient,'se')
    result=(-1)*comp(part,'centx');
  else
    result=comp(part,'centx');
  end
```

```
elseif strcmp(req,'centy')
  if strcmp(orient,'ne') | strcmp(orient,'nw')
    result=(-1)*comp(part,'centY');
  else
    result=comp(part,'centy');
  end
elseif strcmp(req,'comp')
  result(1,1)=lbeam(hl,vl,ht,vt,orient,'area');
  result(1,2)=lbeam(hl,vl,ht,vt,orient,'circ');
  result(1,3)=lbeam(hl,vl,ht,vt,orient,'centx');
  result(1,4)=lbeam(hl,vl,ht,vt,orient,'centy');
  result(1,5)=lbeam(hl,vl,ht,vt,orient,'ix');
  result(1,6)=lbeam(hl,vl,ht,vt,orient,'iy');
elseif strcmp(req,'draw')
  xcentroid=lbeam(hl,vl,ht,vt,orient,'centx');
  ycentroid=lbeam(hl,vl,ht,vt,orient,'centy');
  X=[0 hl hl vt vt 0  0];
  Y=[0 0  ht ht vl vl 0];
  if strcmp(orient,'ne')
    fill(-X,-Y,'r')
  elseif strcmp(orient,'se')
    fill(-X,Y,'r')
  elseif strcmp(orient,'sw')
    fill(X,Y,'r')
  else
    fill(X,-Y,'r')
  end
  hold on;
  ColA=strvcat('Area','Circumference','Centroid X','Centroid
Y','Ix','Iy');
  ColB=makecol(lbeam(hl,vl,ht,vt,orient,'comp')');
  plot (xcentroid,ycentroid,'ko',0,0,'g*')
  hold off
  axis ('equal')
  titleblock(ColA,ColB)
  expandaxis(5,5,5,0)
else
  disp ('That is not a valid request, try area, circ, Ix, Iy,
centX,')
  disp ('centY, comp.')
  disp ('You must use single quotes')
end
```

obeam.m

```
function [result]=obeam(od,id,req)
%OBEAM Circular tube shape routine.
%    OBEAM(OD,ID,REQUEST)
%
%    Shape described: Circular tube.
%
%    Datum location: Center.
%
%    Input arguments:
%       OD: Outer diameter.
%       ID: Inner diameter.
%
%    Requests: (Must be in single quotes)
%       'area':  Area of the shape.
%       'circ':  Circumference of shape.
%       'Ix':    Area moment of inertia about the neutral x axis.
%       'Iy':    Area moment of inertia about the neutral y axis.
%       'centX': Distance from datum to centroid in the x direction.
%       'centY': Distance from datum to centroid in the y direction.
%       'J':     Polar moment of inertia.
%       'comp':  All of the above in a 1x6 matrix.
%       'draw':  Show the shape graphically.
%
%    See also CHANNEL, CIRCLE, COMP, HALFCIRCLE, HORTRAP, HORTRIA, IBEAM,
%       LBEAM, QUARTERCIRCLE, RECTANGLE, RECTUBE, TBEAM, VERTRAP, VERTRIA.

%    Details are to be found in Mastering Mechanics I, Douglas W. Hull,
%    Prentice Hall, 1999

%    Douglas W. Hull, 1999
%    Copyright (c) 1999 by Prentice Hall
%    Version 1.00

req=lower(req);

if (id >= od)
  disp ('The inner diameter must be smaller than the outer diameter')
  return
end
% breaks up the tube into two circles to be used in the
% comp shape routine.
part(1,:)=[circle(od/2,'comp'),0,0,1];
part(2,:)=[circle(id/2,'comp'),0,0,-1]; %the hole
if     strcmp(req,'area')  result=comp(part,'area');
elseif strcmp(req,'circ')  result=comp(part,'circ');
elseif strcmp(req,'Ix')   result=comp(part,'ix');
elseif strcmp(req,'Iy')   result=comp(part,'iy');
elseif strcmp(req,'centx')  result=comp(part,'centx');
elseif strcmp(req,'centy')  result=comp(part,'centy');
elseif strcmp(req,'j')      result=pi/2*(od^4-id^4);
elseif strcmp(req,'comp')
```

```
    result(1,1)=obeam(od,id,'area');
    result(1,2)=obeam(od,id,'circ');
    result(1,3)=obeam(od,id,'centx');
    result(1,4)=obeam(od,id,'centy');
    result(1,5)=obeam(od,id,'ix');
    result(1,6)=obeam(od,id,'iy');
elseif strcmp(req,'draw')
    xcentroid=obeam(od,id,'centx');
    ycentroid=obeam(od,id,'centy');
    angle =[0: DR(1): DR(360)];
    X=[id/2*cos(angle) od/2*cos(angle)];
    Y=[id/2*sin(angle) od/2*sin(angle)];
    fill(X,Y,'r')
    hold on
    ColA=strvcat('Area','Circumference','Centroid X','Centroid
Y','Ix','Iy');
    ColB=makecol(obeam(od,id,'comp')');
    plot (xcentroid,ycentroid,'ko',0,0,'g*')
    hold off;
    axis ('equal')
    titleblock(ColA,ColB)
    expandaxis(5,5,5,0)
else
    disp ('That is not a valid request, try area, circ, Ix, Iy,
centX,')
    disp ('centY, comp.')
    disp ('You must use single quotes')
end
```

rectube.m

```
function [result]=rectube(ob,oh,ib,ih,req)
%RECTUBE Rectangular tube shape routine.
%     RECTUBE(OB,OH,IB,IH,REQUEST)
%
%     Shape described: Rectangular tube.
%
%     Datum location: Bottom left corner.
%
%     Input arguments:
%        OB: Outer base dimension.
%        OH: Outer height dimension.
%        IB: Inner base dimension.
%        IH: Inner height dimension
%
```

```
%     Requests: (Must be in single quotes)
%         'area':  Area of the shape.
%         'circ':  Circumference of shape.
%         'Ix':    Area moment of inertia about the neutral x axis.
%         'Iy':    Area moment of inertia about the neutral y axis.
%         'centX': Distance from datum to centroid in the x direction.
%         'centY': Distance from datum to centroid in the y direction.
%         'comp':  All of the above in a 1x6 matrix.
%         'draw':  Show the shape graphically.
%
%     See also CHANNEL, CIRCLE, COMP, HALFCIRCLE, HORTRAP, HORTRIA, IBEAM,
%         LBEAM, OBEAM, QUARTERCIRCLE, RECTANGLE, TBEAM, VERTRAP, VERTRIA.

%     Details are to be found in Mastering Mechanics I, Douglas W. Hull,
%     Prentice Hall, 1999

%     Douglas W. Hull, 1999
%     Copyright (c) 1999 by Prentice Hall
%     Version 1.00

req=lower(req);

if (ib >= ob)
  disp ('The inner base must be smaller than the outer base')
  return
end
if (ih >= oh)
  disp ('The inner height must be smaller than the outer height')
  return
end
% breaks up the tube into two rectangles to be used in the
% comp shape routine.
part(1,:)=[rectangle(ob,oh,'comp'),0,0,1];
part(2,:)=[rectangle(ib,ih,'comp'),(ob-ib)/2,(oh-ih)/2,-1];
 %the hole
if      strcmp(req,'area')   result=comp(part,'area');
elseif strcmp(req,'circ')   result=comp(part,'circ');
elseif strcmp(req,'ix')   result=comp(part,'ix');
elseif strcmp(req,'iy')   result=comp(part,'iy');
elseif strcmp(req,'centx')   result=comp(part,'centx');
elseif strcmp(req,'centy')   result=comp(part,'centy');
elseif strcmp(req,'comp')
  result(1,1)=rectube(ob,oh,ib,ih,'area');
  result(1,2)=rectube(ob,oh,ib,ih,'circ');
  result(1,3)=rectube(ob,oh,ib,ih,'centx');
  result(1,4)=rectube(ob,oh,ib,ih,'centy');
  result(1,5)=rectube(ob,oh,ib,ih,'ix');
  result(1,6)=rectube(ob,oh,ib,ih,'iy');
```

```
elseif strcmp(req,'draw')
  xcentroid=rectube(ob,oh,ib,ih,'centx');
  ycentroid=rectube(ob,oh,ib,ih,'centy');
  X=[0 ob ob 0 0 (ob-ib)/2 (ob-ib)/2+ib (ob-ib)/2+ib (ob-ib)/2 (ob-
ib)/2 0];
  Y=[0 0 oh oh 0 (oh-ih)/2 (oh-ih)/2 (oh-ih)/2+ih (oh-ih)/2+ih (oh-
ih)/2 0];
  fill(X,Y,'r')
  hold on;
  ColA=strvcat('Area','Circumference','Centroid X','Centroid
Y','Ix','Iy');
  ColB=makecol(rectube(ob,oh,ib,ih,'comp')');
  plot (xcentroid,ycentroid,'ko',0,0,'g*')
  hold off
  axis ('equal')
  titleblock(ColA,ColB)
  expandaxis(5,5,5,0)
else
  disp ('That is not a valid request, try area, circ, Ix, Iy,
centX,')
  disp ('centY, comp.')
  disp ('You must use single quotes')
end
```

channel.m

```
function [result]=channel(b,h,bt,lt,orient,req)
%CHANNEL U-shape shape routine.
%    CHANNEL(BASE,HEIGHT,BT,LT,ORIENT,REQUEST)
%
%    Shape described: A U-shaped channel.
%
%    Datum location: Bottom left corner.
%
%    Input arguments:
%      BASE: Distance between the vertical sides.
%      HEIGHT: Distance between the horizontal sides.
%      BT: Base thickness.
%      LT: Leg thickness.
%      ORIENT: Either 'n','e','s', or 'w'.
%
%    Requests: (Must be in single quotes)
%      'area':  Area of the shape.
%      'circ':  Circumference of shape.
%      'Ix':    Area moment of inertia about the neutral x axis.
%      'Iy':    Area moment of inertia about the neutral y axis.
%      'centX': Distance from datum to centroid in the x direction.
%      'centY': Distance from datum to centroid in the y direction.
%      'comp':  All of the above in a 1x6 matrix.
%      'draw':  Show the shape graphically.
%
```

```
%    See also CIRCLE, COMP, HALFCIRCLE, HORTRAP, HORTRIA, IBEAM, LBEAM,
%     OBEAM, QUARTERCIRCLE, RECTANGLE, RECTUBE, TBEAM, VERTRAP, VERTRIA.

%   Details are to be found in Mastering Mechanics I, Douglas W. Hull,
%   Prentice Hall, 1999

%   Douglas W. Hull, 1999
%   Copyright (c) 1999 by Prentice Hall
%   Version 1.00
req=lower(req);
orient=lower(orient);

if (bt >= h)
  disp ('The height must be greater than the base thickness')
  return
end
if (lt*2 >= b)
  disp ('The base must be bigger than twice the leg thickness')
  return
end
% breaks up the channel into two rectangles to be used in the
% comp shape routine.
part(1,:)=[rectangle(b,h,'comp'),0,0,1];
part(2,:)=[rectangle((b-2*lt),(h-bt),'comp'),lt,bt,-1]; %the hole
if     strcmp(req,'area')  result=comp(part,'area');
elseif strcmp(req,'circ')  result=comp(part,'circ')-(2*(b-2*lt));
elseif strcmp(req,'ix')
  if strcmp(orient,'e')  result=comp(part,'iy');
  elseif strcmp(orient,'s')  result=comp(part,'ix');
  elseif strcmp(orient,'w')  result=comp(part,'iy');
  else  result=comp(part,'ix');
  end
elseif strcmp(req,'iy')
  if strcmp(orient,'e')  result=comp(part,'ix');
  elseif strcmp(orient,'s')  result=comp(part,'iy');
  elseif strcmp(orient,'w')  result=comp(part,'ix');
  else  result=comp(part,'iy');
  end
elseif strcmp(req,'centx')
  if strcmp(orient,'e')  result=comp(part,'centy');
  elseif strcmp(orient,'s')  result=comp(part,'centx');
  elseif strcmp(orient,'w')  result=h-comp(part,'centy');
  else  result=comp(part,'centx');
  end
elseif strcmp(req,'centy')
  if strcmp(orient,'e')  result=comp(part,'centx');
  elseif strcmp(orient,'s')  result=h-comp(part,'centy');
  elseif strcmp(orient,'w')  result=comp(part,'centx');
  else  result=comp(part,'centy');
  end
```

```
elseif strcmp(req,'comp')
  result(1,1)=channel(b,h,bt,lt,orient,'area');
  result(1,2)=channel(b,h,bt,lt,orient,'circ');
  result(1,3)=channel(b,h,bt,lt,orient,'centx');
  result(1,4)=channel(b,h,bt,lt,orient,'centy');
  result(1,5)=channel(b,h,bt,lt,orient,'ix');
  result(1,6)=channel(b,h,bt,lt,orient,'iy');
elseif strcmp(req,'draw')
  xcentroid=channel(b,h,bt,lt,orient,'centx');
  ycentroid=channel(b,h,bt,lt,orient,'centy');
  X=[0 b b (b-lt) (b-lt) lt lt 0 0];
  Y=[0 0 h h bt bt h h 0];
  if strcmp(orient,'n')
    fill(X,Y,'r')
  elseif strcmp(orient,'e')
    fill(Y,X,'r')
  elseif strcmp(orient,'s')
    fill(X,h-Y,'r')
  else
    fill(h-Y,X,'r')
  end
  hold on;
  ColA=strvcat('Area','Circumference','Centroid X','Centroid
Y','Ix','Iy');
  ColB=makecol(channel(b,h,bt,lt,orient,'comp')');
  plot (xcentroid,ycentroid,'ko',0,0,'g*')
  hold off;
  axis ('equal')
  titleblock(ColA,ColB)
  expandaxis(5,5,5,0)
else
  disp ('That is not a valid request, try area, circ, Ix, Iy,
centX,')
  disp ('centY, comp.')
  disp ('You must use single quotes')
end
```

---

ibeam.m

```
function [result]=ibeam(b,h,bt,wt,orient,req)
%IBEAM I-beam shape routine.
%    IBEAM(BASE,HEIGHT,BT,WT,ORIENT,REQUEST)
%
%    Shape described: I-beam.
%
%    Datum location: Bottom left corner.
%
```

```
%    Input arguments:
%      BASE: Distance between the vertical sides.
%      HEIGHT: Distance between the horizontal sides.
%      BT: Base thickness.
%      WT: Web thickness.
%      ORIENT: Either 'I' or 'H'.
%
%    Requests: (Must be in single quotes)
%      'area':  Area of the shape.
%      'circ':  Circumference of shape.
%      'Ix':    Area moment of inertia about the neutral x axis.
%      'Iy':    Area moment of inertia about the neutral y axis.
%      'centX': Distance from datum to centroid in the x direction.
%      'centY': Distance from datum to centroid in the y direction.
%      'comp':  All of the above in a 1x6 matrix.
%      'draw':  Show the shape graphically.
%
%    See also CHANNEL, CIRCLE, COMP, HALFCIRCLE, HORTRAP, HORTRIA, LBEAM,
%      OBEAM, QUARTERCIRCLE, RECTANGLE, RECTUBE, TBEAM, VERTRAP, VERTRIA.
%    Details are to be found in Mastering Mechanics I, Douglas W. Hull,
%    Prentice Hall, 1999

%    Douglas W. Hull, 1999
%    Copyright (c) 1999 by Prentice Hall
%    Version 1.00
req=lower(req);
orient=upper(orient);

if (2*bt >= h)
  disp ('The height must be greater than twice the base thickness')
  return
end
if (wt >= b)
  disp ('The base must be bigger than the web thickness')
  return
end

if ~(strcmp(orient,'I') | strcmp(orient,'H'))
  disp ('Improper orientation')
  return
end
% breaks up the channel into two rectangles to be used in the
% comp shape routine.
part(1,:)=[rectangle(b,bt,'comp'),0,0,1];
part(2,:)=[rectangle(b,bt,'comp'),0,h-bt,1];
part(3,:)=[rectangle(wt,(h-2*bt),'comp'),((b-wt)/2),bt,1];
if     strcmp(req,'area')  result=comp(part,'area');
elseif strcmp(req,'circ')  result=comp(part,'circ')- 4*wt;
elseif strcmp(req,'ix')
  if strcmp(orient,'I')  result=comp(part,'ix');
  else  result=comp(part,'iy');
  end
```

```
elseif strcmp(req,'iy')
  if strcmp(orient,'I')  result=comp(part,'iy');
  else  result=comp(part,'Ix');
  end
elseif strcmp(req,'centx')
  if strcmp(orient,'I')  result=comp(part,'centx');
  else  result=comp(part,'centy');
  end
elseif strcmp(req,'centy')
  if strcmp(orient,'I')  result=comp(part,'centy');
  else  result=comp(part,'centX');
  end
elseif strcmp(req,'comp')
  result(1,1)=ibeam(b,h,bt,wt,orient,'area');
  result(1,2)=ibeam(b,h,bt,wt,orient,'circ');
  result(1,3)=ibeam(b,h,bt,wt,orient,'centx');
  result(1,4)=ibeam(b,h,bt,wt,orient,'centy');
  result(1,5)=ibeam(b,h,bt,wt,orient,'ix');
  result(1,6)=ibeam(b,h,bt,wt,orient,'iy');
elseif strcmp(req,'draw')
  xcentroid=ibeam(b,h,bt,wt,orient,'centx');
  ycentroid=ibeam(b,h,bt,wt,orient,'centy');
  X=[0 b b  (b+wt)/2  (b+wt)/2 b   b 0 0   (b-wt)/2 (b-wt)/2 0  0];
  Y=[0 0 bt bt       h-bt  h-bt h h h-bt h-bt     bt       bt 0];
  if strcmp(orient,'I')
      fill(X,Y,'r')
  else
      fill(Y,X,'r')
  end
  hold on;
  ColA=strvcat('Area','Circumference','Centroid X','Centroid
Y','Ix','Iy');
  ColB=makecol(ibeam(b,h,bt,wt,orient,'comp')');
  plot (xcentroid,ycentroid,'ko',0,0,'g*')
  hold off
  axis ('equal')
  titleblock(ColA,ColB)
  expandaxis(5,5,5,0)
else
  disp ('That is not a valid request, try area, circ, Ix, Iy,
centX,')
  disp ('centY, comp.')
  disp ('You must use single quotes')
end
```

```
function [result]=tbeam(b,h,bt,wt,orient,req)
%TBEAM T-beam shape routine.
%    TBEAM(BASE,HEIGHT,BT,WT,ORIENT,REQUEST)
%
%    Shape described: T-beam.
%
%    Datum location: Bottom left corner.
%
%    Input arguments:
%      BASE: Horizontal dimension.
%      HEIGHT: Verical dimension.
%      BT: Base thickness.
%      WT: Web thickness.
%      ORIENT: Either 'n','e','s' or 'w'.
%
%    Requests: (Must be in single quotes)
%      'area':  Area of the shape.
%      'circ':  Circumfrence of shape.
%      'Ix':    Area moment of inertia about the neutral x axis.
%      'Iy':    Area moment of inertia about the neutral y axis.
%      'centX': Distance from datum to centroid in the x direction.
%      'centY': Distance from datum to centroid in the y direction.
%      'comp':  All of the above in a 1x6 matrix.
%      'draw':  Show the shape graphically.
%
%    See also CHANNEL, CIRCLE, COMP, HALFCIRCLE, HORTRAP, HORTRIA, IBEAM,
%      LBEAM, OBEAM, QUARTERCIRCLE, RECTANGLE, RECTUBE, VERTRAP, VERTRIA.

%    Details are to be found in Mastering Mechanics I, Douglas W. Hull,
%    Prentice Hall, 1999

%    Douglas W. Hull, 1999
%    Copyright (c) 1999 by Prentice Hall
%    Version 1.00

req=lower(req);
orient=lower(orient);

if (bt >= h)
  disp ('The height must be greater than twice the base thickness')
  return
end
if (wt >= b)
  disp ('The base must be bigger than the web thickness')
  return
```

```
end
if ~(strcmp(orient,'n') | strcmp(orient,'e') | ...
  strcmp(orient,'s') | strcmp(orient,'w'))
  disp('Invalid orientation')
  return
end
% breaks up the channel into two rectangles to be used in the
% comp shape routine.
part(1,:)=[rectangle(b,bt,'comp'),0,h-bt,1];
part(2,:)=[rectangle(wt,(h-bt),'comp'),((b-wt)/2),0,1];
if     strcmp(req,'area')  result=comp(part,'area');
elseif strcmp(req,'circ')  result=comp(part,'circ')- 2*wt;
elseif strcmp(req,'ix')
  if strcmp(orient,'n') | strcmp(orient,'s')
result=comp(part,'ix');
  else  result=comp(part,'iy');
  end
elseif strcmp(req,'iy')
  if strcmp(orient,'n') | strcmp(orient,'s')
result=comp(part,'iy');
  else  result=comp(part,'ix');
  end
elseif strcmp(req,'centx')
  if strcmp(orient,'n') | strcmp(orient,'s')
    result=comp(part,'centx');
  elseif strcmp(orient,'e')
    result=comp(part,'centy');
  else
    result=h-comp(part,'centy');
  end
elseif strcmp(req,'centy')
  if strcmp(orient,'n')
    result=comp(part,'centy');
  elseif strcmp(orient,'s')
    result=h-comp(part,'centy')
  else  result=comp(part,'centx');
  end
elseif strcmp(req,'comp')
  result(1,1)=tbeam(b,h,bt,wt,orient,'area');
  result(1,2)=tbeam(b,h,bt,wt,orient,'circ');
  result(1,3)=tbeam(b,h,bt,wt,orient,'centx');
  result(1,4)=tbeam(b,h,bt,wt,orient,'centy');
  result(1,5)=tbeam(b,h,bt,wt,orient,'ix');
  result(1,6)=tbeam(b,h,bt,wt,orient,'iy');
```

```
elseif strcmp(req,'draw')
   xcentroid=tbeam(b,h,bt,wt,orient,'centx');
   ycentroid=tbeam(b,h,bt,wt,orient,'centy');
   X=[(b-wt)/2 (b+wt)/2 (b+wt)/2 b b 0 0 (b-wt)/2 (b-wt)/2];
   Y=[0 0 h-bt h-bt h h h-bt h-bt 0];
   if strcmp(orient,'n')
      fill(X,Y,'r')
   elseif strcmp(orient,'e')
      fill(Y,X,'r')
   elseif strcmp(orient,'s')
      fill(X,b-Y,'r')
   else
      fill(h-Y,X,'r')
   end
   hold on;
   ColA=strvcat('Area','Circumference','Centroid X','Centroid
Y','Ix','Iy');
   ColB=makecol(tbeam(b,h,bt,wt,orient,'comp')');
   plot (xcentroid,ycentroid,'ko',0,0,'g*')
   hold off
   axis ('equal')
   titleblock(ColA,ColB)
   expandaxis(5,5,5,0)
else
   disp ('That is not a valid request, try area, circ, Ix, Iy,
centX,')
   disp ('centY, comp.')
   disp ('You must use single quotes')
end
```

hortrap.m

```
function [result]=hortrap(b,h,a,p,req)
%HORTRAP Horizontal trapezoid shape routine.
%    HORTRAP(BASE,HEIGHT,A,P,REQUEST)
%
%    Shape described: Trapezoid with a horizontal base.
%
%    Datum location: Leftmost point of base.
%
%    Input arguments:
%     BASE: Length of base (base is the larger of the horizontal sides).
%     HEIGHT: Distance between the horizontal sides can be negative.
%      A: Length of the shorter of the horizontal sides.
%      P: Horizontal distance from the datum to lower edge of small edge.
%
```

```
%    Requests: (Must be in single quotes)
%      'area':  Area of the shape.
%      'circ':  Circumference of shape.
%      'Ix':    Area moment of inertia about the neutral x axis.
%      'Iy':    Area moment of inertia about the neutral y axis.
%      'centX': Distance from datum to centroid in the x direction.
%      'centY': Distance from datum to centroid in the y direction.
%      'comp':  All of the above in a 1x6 matrix.
%      'draw':  Show the shape graphically.
%
%    See also CHANNEL, CIRCLE, COMP, HALFCIRCLE, HORTRIA, IBEAM, LBEAM,
%       OBEAM, QUARTERCIRCLE, RECTANGLE, RECTUBE, TBEAM, VERTRAP, VERTRIA.

%    Details are to be found in Mastering Mechanics I, Douglas W. Hull,
%    Prentice Hall, 1998

%    Douglas W. Hull, 1998
%    Copyright (c) 1998-99 by Prentice Hall
%    Version 1.00

req=lower(req);

if (a > b)
  disp ('"B" value may not be larger than "B" value.  Remember to use ')
  disp ('proper sign convention for "H"')
  return
end
if (p+a > b) | (p < 0)
  disp ('"A" side of trapezoid may not extend beyond the boundary ')
  disp ('of the base')
  return
end

Origh=h;
h=abs(h);
IsPos= (Origh==h)*2-1;
% breaks up the trapezoid into two triangles and a rectangle to be
used in
% the comp shape routine.
part(1,:)=[hortria(p,h,p,'comp'),0,0,1];
part(2,:)=[rectangle(a,h,'comp'),p,0,1];
part(3,:)=[hortria((b-p-a),h,0,'comp'),(p+a),0,1];
if     strcmp(req,'area')   result=h*(a+b)/2;
elseif strcmp(req,'circ')   result=sqrt(p^2+h^2)+a+sqrt(h^2+(b-a-
p)^2)+b;
elseif strcmp(req,'ix')  result=comp(part,'Ix');
elseif strcmp(req,'iy')  result=comp(part,'Iy');
elseif strcmp(req,'centx')  result=comp(part,'centX');
elseif strcmp(req,'centy')  result=comp(part,'centY')*IsPos;
elseif strcmp(req,'comp')
```

```
   result(1,1)=hortrap(b,Origh,a,p,'area');
   result(1,2)=hortrap(b,Origh,a,p,'circ');
   result(1,3)=hortrap(b,Origh,a,p,'centx');
   result(1,4)=hortrap(b,Origh,a,p,'centy');
   result(1,5)=hortrap(b,Origh,a,p,'ix');
   result(1,6)=hortrap(b,Origh,a,p,'iy');
elseif strcmp(req,'draw')
   xcentroid=hortrap(b,Origh,a,p,'centx');
   ycentroid=hortrap(b,Origh,a,p,'centy');
   X=[0 b p+a  p 0];
   Y=[0 0 Origh Origh 0];
   fill(X,Y,'r')
   hold on;
   ColA=strvcat('Area','Circumference','Centroid X','Centroid
Y','Ix','Iy');
   ColB=makecol(hortrap(b,Origh,a,p,'comp')');
   plot (xcentroid,ycentroid,'ko',0,0,'g*')
   hold off
   axis ('equal')
   titleblock(ColA,ColB)
   expandaxis(5,5,5,0)
else
   disp ('That is not a valid request, try area, circ, Ix, Iy,
centX,')
   disp ('centY, comp.')
   disp ('You must use single quotes')
end
```

vertrap.m

```
function [result]=vertrap(b,h,a,p,req)
%VERTRAP Horizontal trapezoid shape routine.
%    VERTRAP(BASE,HEIGHT,A,P,REQUEST)
%
%    Shape described: Trapezoid with a vertical base.
%
%    Datum location: Lowermost point of base.
%
%    Input arguments:
%      base: Length of base (base is the larger of the vertical
sides).
%      height: Distance between the vertical sides, may be negative.
%      a: Length of the shorter of the vertical sides.
%      p: Vertical distance from the datum to lower edge of small edge.
```

```
%
%    Requests: (Must be in single quotes)
%       'area':  Area of the shape.
%       'circ':  CircumfERENce of shape.
%       'Ix':    Area moment of inertia about the neutral x axis.
%       'Iy':    Area moment of inertia about the neutral y axis.
%       'centX': Distance from datum to centroid in the x direction.
%       'centY': Distance from datum to centroid in the y direction.
%       'comp':  All of the above in a 1x6 matrix.
%       'draw':  Show the shape graphically.
%
%    See also CHANNEL, CIRCLE, COMP, HALFCIRCLE, HORTRAP, HORTRIA, IBEAM,
%       LBEAM, OBEAM, QUARTERCIRCLE, RECTANGLE, RECTUBE, TBEAM, VERTRIA.

%    Details are to be found in Mastering Mechanics I, Douglas W. Hull,
%    Prentice Hall, 1998

%    Douglas W. Hull, 1998
%    Copyright (c) 1998-99 by Prentice Hall
%    Version 1.00

req=lower(req);

if (a > b)
  disp ('"A" value may not be larger than "B" value.  Remember to
use ')
  disp ('proper sign convention for "H"')
  return
end
if (p+a > b) | (p < 0)
  disp ('"A" side of trapezoid may not extend beyond the boundary ')
  disp ('of the base')
  return
end

Origh=h;
h=abs(h);
IsPos= (Origh==h)*2-1;
% breaks up the trapezoid into two triangles and a rectangle to be
used in
% the comp shape routine.
part(1,:)=[vertria(b-p-a,h,0,'comp'),0,(a+p),1];
part(2,:)=[rectangle(h,a,'comp'),0,p,1];
part(3,:)=[vertria(p,h,p,'comp'),0,0,1];
```

```
if       strcmp(req,'area')    result=h*(a+b)/2;
elseif strcmp(req,'circ')    result=sqrt(p^2+h^2)+a+sqrt(h^2+(b-a-
p)^2)+b;
elseif strcmp(req,'ix')   result=comp(part,'ix');
elseif strcmp(req,'iy')   result=comp(part,'iy');
elseif strcmp(req,'centx')   result=comp(part,'centx')*IsPos;
elseif strcmp(req,'centy')   result=comp(part,'centy');
elseif strcmp(req,'comp')
   result(1,1)=vertrap(b,Origh,a,p,'area');
   result(1,2)=vertrap(b,Origh,a,p,'circ');
   result(1,3)=vertrap(b,Origh,a,p,'centx');
   result(1,4)=vertrap(b,Origh,a,p,'centx');
   result(1,5)=vertrap(b,Origh,a,p,'ix');
   result(1,6)=vertrap(b,Origh,a,p,'iy');
elseif strcmp(req,'draw')
   xcentroid=vertrap(b,Origh,a,p,'centx');
   ycentroid=vertrap(b,Origh,a,p,'centy');
   X=[0 Origh Origh 0 0];
   Y=[0 p p+a b 0];
   fill(X,Y,'r')
   hold on;
   ColA=strvcat('Area','Circumference','Centroid X','Centroid
Y','Ix','Iy');
   ColB=makecol(vertrap(b,Origh,a,p,'comp')');
   plot (xcentroid,ycentroid,'ko',0,0,'g*')
   hold off
   axis ('equal')
   titleblock(ColA,ColB)
   expandaxis(5,5,5,0)
else
   disp ('That is not a valid request, try area, circ, Ix, Iy,
centX,')
   disp ('centY, comp.')
   disp ('You must use single quotes')
end
```

8.5   Output

These routines can give much of the engineering information needed from these shapes. The following examples will show how to describe the standard shapes and how to find information about them.

For each of the following shapes, find the

- Area
- Circumference
- Centroid
- I_x and I_y

```
>>Aarea=lbeam(8,5,1,2,'nw','area')
Aarea =
        16
>>Acirc=lbeam(8,5,1,2,'nw','circ')
Acirc =
        26
>>Acentx=lbeam(8,5,1,2,'nw','centX
')
AcentX =
    2.5000
>>Acenty=lbeam(8,5,1,2,'nw','centY
')
Acenty =
    -1.7500
>>Aix=lbeam(8,5,1,2,'nw','Ix')
Aix =
    36.3333
>>Aiy=lbeam(8,5,1,2,'nw','Iy')
Aiy =
    81.3333
>>lbeam(8,5,1,2,'nw','draw')
```

a.)

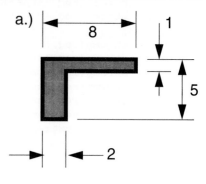

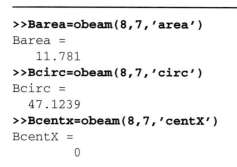

```
>>Barea=obeam(8,7,'area')
Barea =
    11.781
>>Bcirc=obeam(8,7,'circ')
Bcirc =
    47.1239
>>Bcentx=obeam(8,7,'centX')
BcentX =
        0
```

b.)

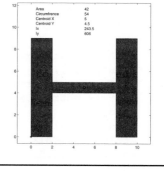

inner diameter: 7
outer diameter: 8

```
>>Bcenty=obeam(8,7,'centY')
Bcenty =
        0
>>Bix=obeam(8,7,'Ix')
Bix =
  83.2031
>>Biy=obeam(8,7,'Iy')
Biy =
  83.2031
>>obeam(8,7,'draw')
```

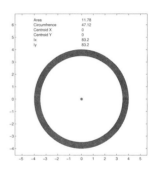

```
>>Carea=rectube(9,5,5,3,'area')
Carea =
        30
>>Ccirc=rectube(9,5,5,3,'circ')
Ccirc =
        44
>>Ccentx=rectube(9,5,5,3,'centX')
CcentX =
  4.5000
>>Ccenty=rectube(9,5,5,3,'centY')
Ccenty =
  2.5000
>>Cix=rectube(9,5,5,3,'Ix')
Cix =
  82.5000
>>Ciy=rectube(9,5,5,3,'Iy')
Ciy =
 272.5000
>>rectube(9,5,5,3,'draw')
```

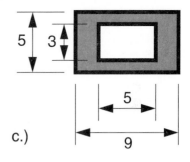

c.)

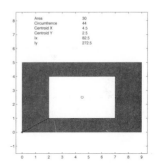

```
>>Darea=channel(9,5,2,1,'s','area'
)
Darea =
      24
>>Dcirc=channel(9,5,2,1,'s','circ'
)
Dcirc =
      34
>>Dcentx=channel(9,5,2,1,'s','cent
X')
DcentX =
     4.5
>>Dcenty=channel(9,5,2,1,'s','centY')
Dcenty =
    3.375
>>Dix=channel(9,5,2,1,'s','Ix')
Dix =
   38.625
>>Diy=channel(9,5,2,1,'s','Iy')
Diy =
      218
>>channel(9,5,2,1,'s','draw')
```

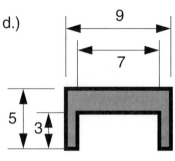

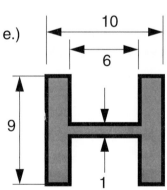

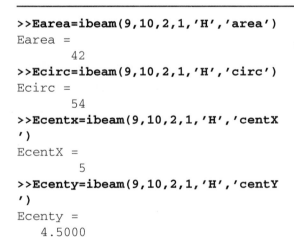

```
>>Earea=ibeam(9,10,2,1,'H','area')
Earea =
      42
>>Ecirc=ibeam(9,10,2,1,'H','circ')
Ecirc =
      54
>>Ecentx=ibeam(9,10,2,1,'H','centX
')
EcentX =
       5
>>Ecenty=ibeam(9,10,2,1,'H','centY
')
Ecenty =
    4.5000
```

```
>>Eix=ibeam(9,10,2,1,'H','Ix')
Eix =
 243.5000
>>Eiy=ibeam(9,10,2,1,'H','Iy')
Eiy =
      606
>>ibeam(9,10,2,1,'H','draw')
```

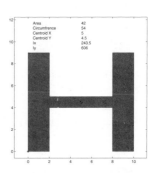

```
>>Farea=hortrap(10,-4,3,6,'area')
Farea =
       26
>>Fcirc=hortrap(10,-4,3,6,'circ')
Fcirc =
   24.3342
>>Fcentx=hortrap(10,-
4,3,6,'centX')
FcentX =
   6.0256
>>Fcenty=hortrap(10,-
4,3,6,'centY')
Fcenty =
   -1.6410
>>Fix=hortrap(10,-4,3,6,'Ix')
Fix =
   31.3162
>>Fiy=hortrap(10,-4,3,6,'Iy')
Fiy =
  130.3162
>>hortrap(10,-4,3,6,'draw')
```

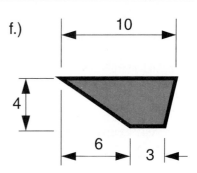

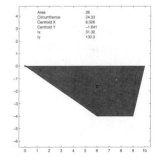

111

```
>>Garea=vertrap(10,4,5,4,'area')
Garea =
        30
>>Gcirc=vertrap(10,4,5,4,'circ')
Gcirc =
     24.78
>>Gcentx=vertrap(10,4,5,4,'centX')
GcentX =
     1.7778
>>Gcenty=vertrap(10,4,5,4,'centY')
Gcenty =
     5.6667
>>Gix=vertrap(10,4,5,4,'Ix')
Gix =
   161.6667
>>Giy=vertrap(10,4,5,4,'Iy')
Giy =
    38.5185
>>vertrap(10,4,5,4,'draw')
```

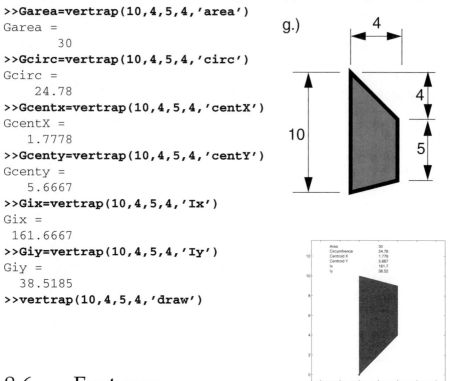

8.6   Features

These functions all make use of the *comp.m* function. That function does all of the real work. These functions can all be thought of as "wrappers", programs that wrap around another program to make the usage easier. As will be seen in the next chapter, these shapes could all be calculated without creating a function dedicated to it's solution. These dedicated functions are much easier to use, for the one time expense of their creation.

8.7   Summary

Required argument

[answer]=lbeam(*hl, vl, ht, vt, orientation, 'keyword'*)

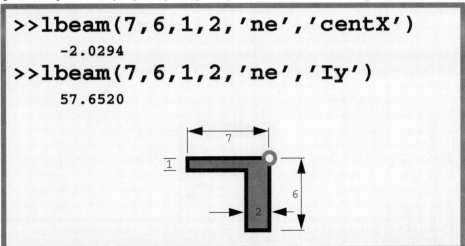

```
>>lbeam(7,6,1,2,'ne','centX')
    -2.0294
>>lbeam(7,6,1,2,'ne','Iy')
    57.6520
```

[answer]=obeam(*od,id,'keyword'*)

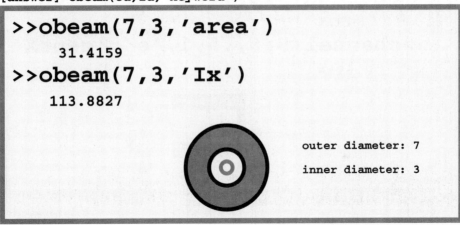

```
>>obeam(7,3,'area')
    31.4159
>>obeam(7,3,'Ix')
    113.8827
```

outer diameter: 7

inner diameter: 3

113

`[answer]=rectube(`*`ob,oh,ib,ih,`*`'keyword')`

```
>>rectube(8,4,6,2,'Ix')
     38.6667
>>rectube(8,4,6,2,'Iy')
    134.6667
```

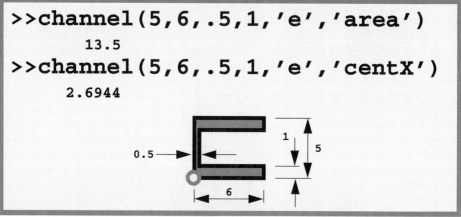

`[answer]=channel(`*`b,h,bt,lt,orientation,`*`'keyword')`

```
>>channel(5,6,.5,1,'e','area')
     13.5
>>channel(5,6,.5,1,'e','centX')
    2.6944
```

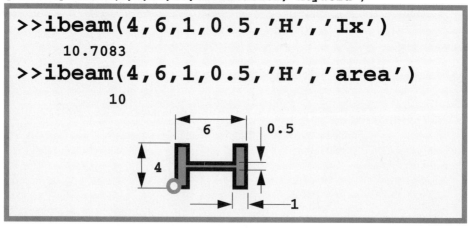

`[answer]=ibeam(`*`b,h,bt,wt,orientation,`*`'keyword')`

```
>>ibeam(4,6,1,0.5,'H','Ix')
    10.7083
>>ibeam(4,6,1,0.5,'H','area')
    10
```

114

[answer]=ibeam(b,h,bt,wt,orientation,'keyword')

```
>>tbeam(4,6,1,0.5,'e','Ix')
     5.3854
>>tbeam(4,6,1,0.5,'e','area')
     6.5000
```

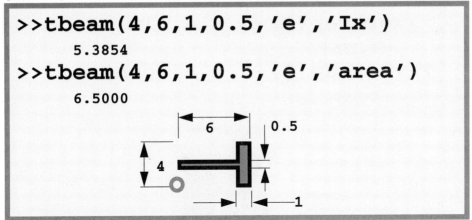

[answer]=hortrap(b,h,a,p,'keyword')

```
>>hortrap(9,3,1,6,'Ix')
     8.8500
>>hortrap(9,3,1,6,'centX')
     5.2333
```

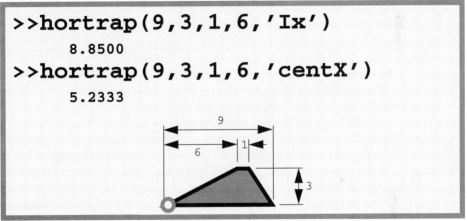

`[answer]=vertrap(b,h,a,p,'keyword')`

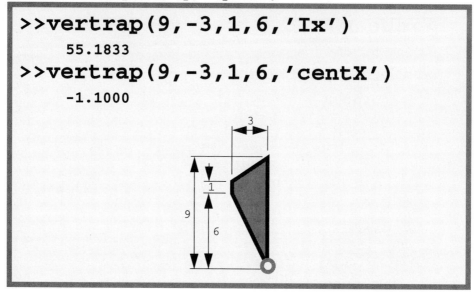

```
>>vertrap(9,-3,1,6,'Ix')
    55.1833
>>vertrap(9,-3,1,6,'centX')
    -1.1000
```

The "keyword" argument for these functions must include the single quotes and can be chosen from this list:

- 'area' the area of the shape.
- 'circ' the length of the perimeter of the shape.
- 'centX' the distance from the datum to the centroid along the x axis.
- 'centY' the distance from the datum to the centroid along the y axis.
- 'Ix' the area moment of inertia about the x axis through the centroid.
- 'Iy' the area moment of inertia about the y axis through the centroid.
- 'comp' all of the above into a 1x6 matrix. This is useful for composite shapes.
- 'draw' the graphical representation of the shape.

9

Geometry of Composite Shapes

9.1   Introduction

The standard composite shapes explored in the previous chapter are useful, but are limited in scope. The routine of this chapter will work for any combination of primitives imaginable. The set-up time for each new shape is longer, but the added flexibility is at times essential.

9.2   Problem

Given any composition of primitive shapes, find the following engineering values:
- Area
- Circumference
- Area moments of inertia
- Centroid

9.3   Theory

The primitive shapes can be combined in an infinite number of ways, but the same rules apply to their combinations no matter the configuration. Those rules will be outlined in the following text.

Area: The area of the composite is the sum of the nonoverlapping subareas. The primitive parts are not allowed to overlap, therefore the area from each of the subparts may simply be summed together. There is the added complication of holes in parts. These are simply subtracted from the total rather than added.

Circumference: As a general rule, this feature will **not** give the correct answer. It will only sum the circumferences of the subparts. This of course will often give erroneous results because shared borders of subparts are internal and should not be counted at all. In the simple method of summing, the circumferences will actually be counted twice. Depending on the specific situation, this can be remedied by subtracting off the appropriate internal distance values:

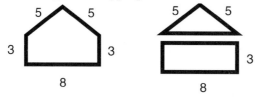

actual circumference: 24   sum of circumferences: 40

Centroid: The centroid of a composite part is found with a weighted average of all the subparts. For this example, the X distance to the centroid can be found.

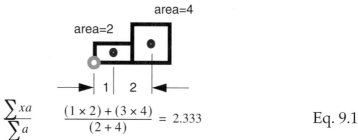

$$\frac{\sum xa}{\sum a} \qquad \frac{(1 \times 2) + (3 \times 4)}{(2 + 4)} = 2.333 \qquad \text{Eq. 9.1}$$

The answer will be returned as the distance from the datum of the composite shape.

Area Moment of Inertia: The area moment of inertia of the entire shape is found by moving the moment of inertia of each of the subparts to the centroid of the entire shape. The moments of inertia are then summed together. The moments of inertia are moved from the centroid of each subpart to the centroid of the entire shape by use of the parallel axis theorem.

$$I_x = \overline{I}_x + Ad^2 \qquad \text{Eq. 9.2}$$

9.4   Template

```
function [result]=comp(part,req)
%COMP Composite shape routine.
%    COMP(PARTS,REQUEST)
%
%    Shape described: A composite shape.
%
%    Datum location: Arbitrary, all shapes are individually placed.
%
%    Input arguments:
%      PARTS: An N by 9 matrix containing data on the components of the
%         composite shape.
%      PARTS=[area, circ, centX, centY, Ix, Iy, X, Y, SIGN]
%        The first six pieces of information are easily supplied with other
%           shape functions using the "comp" request.
%        X and Y are the the distance from the composite datum to the datum
%           of each individual shape datum.
%        SIGN is either (1) or (-1) with (1) meaning the shape represents
%           material and (-1) meaning the shape represents a hole in the
%           material.
%
%    Requesting circumference is a feature simply for compliance with other
%       shape functions.  The answer arrived at for the circumference is never
%       correct due to shared inner borders.  Some corrections can be done to
%       the answer to find the correct value.
%
%    Requests: (Must be in single quotes)
%      'area':  Area of the shape.
%      'circ':  Circumference of shape.
%      'Ix':    Area moment of inertia about the neutral x axis.
%      'Iy':    Area moment of inertia about the neutral y axis.
%      'centX': Distance from datum to centroid in the x direction.
%      'centY': Distance from datum to centroid in the y direction.
%      'comp':  All of the above in a 1x6 matrix.
%      'draw':  Show the shape graphically.
%
%    See also CHANNEL, CIRCLE, HALFCIRCLE, HORTRAP, HORTRIA, IBEAM, LBEAM,
%       OBEAM, QUARTERCIRCLE, RECTANGLE, RECTUBE, TBEAM, VERTRAP, VERTRIA.

%    Prentice Hall, 1999

%    Douglas W. Hull, 1999
%    Copyright (c) 1999 by Prentice Hall
%    Version 1.00
```

```
req=lower(req);

if      strcmp(req,'area')   result=sum (part(:,1).*part(:,9));
elseif strcmp(req,'circ')    result=sum (part(:,2));
elseif strcmp(req,'ix')
  cent=comp(part,'centy');
  d=cent-(part(:,4)+part(:,8));
  result=sum((part(:,5)+part(:,1).*(d.^2)).*part(:,9));

elseif strcmp(req,'iy')
  cent=comp(part,'centx');
  d=cent-(part(:,3)+part(:,7));
  result=sum((part(:,6)+part(:,1).*(d.^2)).*part(:,9));

elseif strcmp(req,'centx')
  num=sum((part(:,3)+part(:,7)).*part(:,1).*part(:,9));
  den=sum(part(:,1).*part(:,9));
  result= num/den;
elseif strcmp(req,'centy')
  num=sum((part(:,4)+part(:,8)).*part(:,1).*part(:,9));
  den=sum(part(:,1).*part(:,9));
  result= num/den;
elseif strcmp(req,'comp')
  result(1,1)=comp(part,'area');
  result(1,2)=comp(part,'circ');
  result(1,3)=comp(part,'centx');
  result(1,4)=comp(part,'centy');
  result(1,5)=comp(part,'ix');
  result(1,6)=comp(part,'iy');
else
  disp ('That is not a valid request, try area, circ, Ix, Iy,
centX,')
  disp ('centY, comp.')
  disp ('You must use single quotes')
end
```

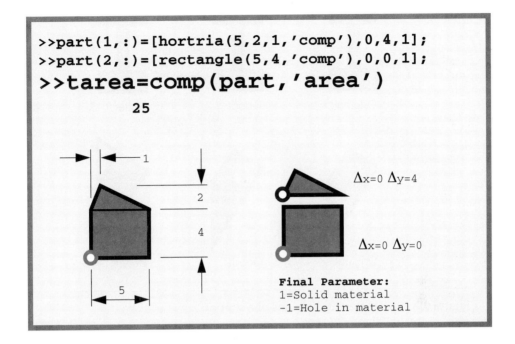

```
>>part(1,:)=[hortria(5,2,1,'comp'),0,4,1];
>>part(2,:)=[rectangle(5,4,'comp'),0,0,1];
>>tarea=comp(part,'area')
          25
```

$\Delta x=0$ $\Delta y=4$

$\Delta x=0$ $\Delta y=0$

Final Parameter:
```
1=Solid material
-1=Hole in material
```

9.5   Output

Getting output directly from *comp.m* requires a bit of preparation work, but it is a powerful function. This is the function that powered the routines *ibeam.m* *channel.m, hortrap.m,* and the other standard composite shapes. The steps in using *comp.m* are:

- Dividing the shape into known shapes
- Closing a datum point
- Locating primitives relative to datum
- Creating the parts matrix
- Invoking the function

Dividing into known shapes: Any division of the shape into nonoverlapping parts will work, however certain divisions will be simpler than others, see Figure 9.1. Remember that any shape that has been defined previously will work. If a shape is an I-beam that has one edge thicker than the other it can be divided into an I-beam and an extra rectangle. This would save a step over using three rectangles.

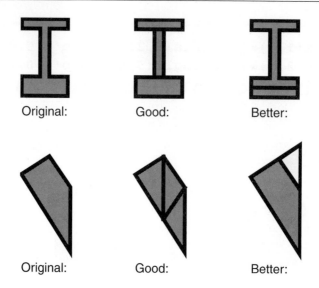

Figure 9.1   Different divisions of a shape.

When dividing shapes, do not be afraid to use negative shape, holes or cut outs if it will simplify the process. As with most things, the fewer steps in the process the better.

Choosing a datum point: The datum choice is entirely arbitrary. Choose it to your own liking according to the nature of the problem. It does not effect the final results.

Locating the Primitives: The location of the datum of each of the primitives must be calculated. By a judicious choice of primitives and datum placement, these measurements should be easy.

Creating the Parts Matrix: This is the first bit of MATLAB code that needs to be written for the shape. A matrix of the following form needs to be created:

$$\begin{bmatrix} \text{area}_1 & \text{circumference}_1 & \text{centroid x}_1 & \text{centroid x}_1 & Ix_1 & Iy_1 & \Delta x_1 & \Delta y_1 & \text{multiplier}_1 \\ \text{area}_2 & \text{circumference}_2 & \text{centroid x}_2 & \text{centroid x}_2 & Ix_2 & Iy_2 & \Delta x_2 & \Delta y_2 & \text{multiplier}_2 \\ \text{area}_3 & \text{circumference}_3 & \text{centroid x}_3 & \text{centroid x}_3 & Ix_3 & Iy_3 & \Delta x_3 & \Delta y_3 & \text{multiplier}_3 \end{bmatrix} \quad \text{Eq. 9.3}$$

At first this looks like an unreasonable amount of data to put together. However a second look at the first six columns shows that they are actually all of the data that is available from the shape functions that have already been designed. The functions allow the user to call for the data one piece at a time, or by using the keyword "comp" all the data is sent at once. Knowing that, the input matrix looks much simpler:

$$\begin{bmatrix} Shape1(a,b,c,'comp') & \Delta x_1 & \Delta y_1 & multiplier_1 \\ Shape2(a,b,c,'comp') & \Delta x_2 & \Delta y_2 & multiplier_2 \\ Shape3(a,b,c,'comp') & \Delta x_3 & \Delta y_3 & multiplier_3 \end{bmatrix}$$

Eq. 9.4

Looking at Equation 9.4 the matrix looks much more manageable. Each row of the matrix is the output from the "comp" keyword of each shape, followed by the distance of the datum of the composite shape to the datum of the shape. Last is the multiplier, either a "1" for material being present or "-1" if there is a hole. Conceivably other multipliers can be used, making the subshape more or less important in the calculation. There seems to be little use for this however. Unity and negative unity should be sufficient for normal use.

a.) Δx: 0 Δy: 6 multiplier: -1
b.) Δx: 0 Δy: 0 multiplier: 1
c.) Δx: 9 Δy: 3 multiplier: 1

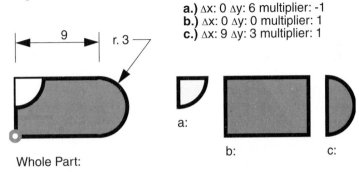

Whole Part:

Figure 9.2   Sample composite part.

To create the parts matrix for this composite part Figure 9.2 do as follows:
```
>>part(1,:)=[quartercircle(3,'se','comp') 0 6 -1];
>>part(2,:)=[rectangle(9,6,'comp') 0 0 1];
>>part(3,:)=[halfcircle(3,'e','comp') 9 3 1];
```
Any of the above lines can be read as: "For the matrix called part, let the first row and any number of columns equal the output from the quarter circle function followed by the three rows 0, 6, and -1"

Invoking the Function: Now that the parts matrix is defined, it may be fed into *comp.m* along with a keyword to designate the requested information.
```
>>area=comp(part,'area')
area =
   61.0686
>>centX=comp(part,'centX')
centX =
    6.2100
```

It was that simple. One line for each of the parts making up the composite shape, and one line for each request!

If the circumference of the part was required, a bit more thinking would be needed. After requesting the circumference, which is the sum of the smaller shapes, some amount will need to be subtracted away. In this example the two flat edges of the quarter circle must be taken away, then the upper left corner of the rectangle along with the entire right edge. The diameter of the semicircle must also be deleted. All told this is eight times the radius that must be subtracted.

```
>>circ=comp(part,'circ')-(8*3)
circ =
   32.1372
```

9.6   Features

Although this function is designed to work handily with the other shape functions used to build the part matrix using the "comp" keyword, it is conceivable that the needed data was found through testing, handbooks or other outside sources. In these cases the matrix could be assembled by whatever means is appropriate.

9.7   Summary

Required argument

```
[answer]=comp([part],'keyword')
```

The part matrix has one row for each subpart that makes up the composite shape. This part matrix is of the form shown in Equation 9.3 or in simpler form in Equation 9.4. The most usual method for constructing the matrix is:

```
part(1,:)=[shape(a,b,c,'comp'), Delta_X, Delta_Y,
Multiplier];
```

Where ΔX and ΔY are the distances from the datum of the composite shape to the datum of the individual shapes. The multiplier is either "1" for solid material or "-1" to represent material that has been cut away. As many rows as are necessary can be added to the matrix to represent all of the individual shapes.

The "keyword" argument for these functions must include the single quotes and can be chosen from this list:

- 'area' the area of the shape.
- 'circ' the length of the perimeter of the shape.
- 'centX' the distance from the datum to the centroid along the x axis.
- 'centY' the distance from the datum to the centroid along the y axis.
- 'Ix' the area moment of inertia about the x axis through the centroid.
- 'Iy' the area moment of inertia about the y axis through the centroid.
- 'comp' all of the above into a 1x6 matrix. This is useful for composite shapes.

10

Average Normal and Shear Stresses

10.1 Introduction

Stress can be thought of as the amount of "distress" that the material is feeling due to the forces that are being applied to it. In it's simplest form, stress is a measure of the force per unit area seen by the material. Because stress is related to the forces on the part and the area of specified cross sections of the part, it can be seen why so much time has been taken to this point to understand the forces in a body and why so much has been done to find the areas of shapes.

10.2 Problem

Given a simple situation, find the average stresses that a part is exposed to. With all of the background tools that have been developed thus far, this should be a matter of putting together pieces of information.

10.3 Theory

The standard assumptions of homogenous, isotropic material will hold in this discussion. The theory is fairly simple:

$$\text{Average Normal Stress} = \sigma_{avg} = \frac{n}{a}$$

$$\text{Average Shear Stress} = \tau_{avg} = \frac{v}{a}$$

Eq. 10.1

Where n is the normal force, v is the shear force, and a is the area.

10.4   Output

Some cases of ways to find the average stress on a plane are shown here in Figure 10.1.

Example A: What is the normal stress at each of the labeled sections?

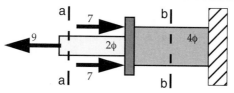

Figure 10.1   Normal stress problem, Example A.

This problem is best solved with the method of sections. Drawing the free body diagram through plane *a-a* and taking everything to the left, by inspection the reaction force at *a-a* is:

```
>>forceaa=[9 0 0 0];
>>normalaa=mag(forceaa,'x');
```

The horizontal length of these members are rather irrelevant so dummy values may simply be entered.

The reaction forces at *b-b* could be done by inspection also, but for the sake of example:

```
>>af=[-9 0 0 0;7 0 5 2;7 0 5 -2];
>>[forcebb momentbb]=reaction(af,[10 0]);
>>normalbb=mag(forcebb,'x');
```

Knowing the normal force at each of the cross sections, it now becomes necessary to find the area at the cross sections.

```
>>areaaa=circle(2,'area');
>>areabb=circle(4,'area');
>>nsaa=normalaa/areaaa
nsaa =
    0.7162
>>nsbb=normalbb/areabb
nsbb =
    0.0995
```

What was the general procedure for this?

1. Find the centroid of the plane of interest.
2. Find the normal and/or shear forces at the centroid of the plane of interest.
3. Convert these forces into a magnitude of the correct direction only, *mag.m.*
4. Find the area of the cross section
5. Divide the force by the area to find the stress.

This example was simple because all the forces were axial and the cross section was already normal to the only forces, meaning that there was no shear force. This process can be applied to more difficult situations.

Example B: What is the normal and shear stress on the shaded plane shown in Figure 10.2?

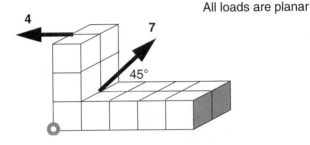

Figure 10.2   Loaded structure, Example B.

CH1001

```
af=[-4 0 0 3;deg2xy([45 7 1 1])];
ycord=rectangle(2,1,'centY'); % Find centroid of section
[force,momment]=reaction(af,[5,ycord]);
fnormal=mag(force,'x');
fshear=mag(force,'y');
area=rectangle(2,1,'area');
normalstress=fnormal/area
```

```
normalstress =
  -0.4749
>>shearstress=fshear/area
shearstress =
  -2.4749
```

By looking at the magnitudes of the stresses the directions of the stress can be figured out. For example a -0.4749 stress on the face would mean a compressive stress because a force in the negative X must have been acting on the face of the block, a force in that direction would compress the block.

This example was more difficult because a shear force existed and the centroid of the cross section had to be calculated. Though the centroid could be seen from inspection, it is not difficult to imagine scenerios where the calculation would be less trivial.

For the problems of this section, the centroid of the cross section does not actually need to be calculated, the reaction forces will be the same anywhere on the face. However, the reaction moment will be different depending on the position on the face. The reaction moment is significant in bending calculations that will often accompany the normal and shear calculations. For this reason it will be necessary to calculate the centroid for a full analysis —it is a good habit to start now.

Example C: What are the average shear and normal forces on the dark shaded face, as illustrated in Figure 10.3?

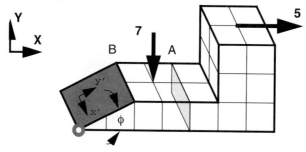

Figure 10.3   Loaded structure, Example C.

CH1002.m

```
af=[5 0 6 3; 0 -7 3 1];
Aarea=rectangle(2,1,'area');
AcentY=rectangle(2,1,'centY');
phi=atan2(1,2);
Barea=csc(phi)*Aarea;
BcentY=csc(phi)*AcentY;
centX=BcentY*cos(phi);
centY=AcentY;
```

Stop for a moment to see if the data generated thus far is reasonable, and to revise the free body diagram, see Figure 10.4.

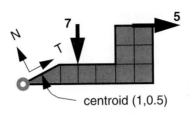

centroid (1,0.5)

Figure 10.4   Revised free body diagram, Example C.

These values seem reasonable. The next step is to find the reaction forces at the centroid of the cross section of interest. Once the reaction forces are found, they can be resolved into the normal and tangent directions.

```
xyreaction=reaction(af,[1,0.5]);
ntreaction=twovector(opp(xyreaction),[phi, phi+DR(90)]);
tangent=ntreaction(1,:);
normal=ntreaction(2,:);
NormalStress=mag(normal)/Barea
```

```
NormalStress =
    1.9000
>>TangentStress=mag(tangent)/Barea
TangentStress =
    0.3000
```

Example D: What is the shear stress on the pin at *a* if it is in double shear as shown in Figure 10.5?

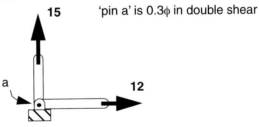

'pin a' is 0.3ϕ in double shear

Figure 10.5   Example D.

Following the general procedure, first the centroid of the plane of interest must be found. This can be done by inspection for a round pin. Next, the normal and shear forces on that plane must be found. The shear force exists only on this pin. Furthermore, the shear force is just the sum of the forces acting on the point.

```
>>af=[0,15,0,0;12,0,0,0];
>>aforce=sumforce(af);
```

The next point of the procedure calls for the magnitude of the force.

```
>>amag=mag(aforce);
```

Then the area that supports the load must be found. The shape routines are useful for this. However, remember the pin is in double shear, so the area of the pin must be doubled.

```
>>area=2*circle(0.3/2,'area');
```

Finally, the magnitude of the force divided by the area supporting the force can be calculated to give the average stress.

```
>>ShearStress=amag/area
ShearStress =
  135.8785
```

10.5   Summary

The procedure for finding the average shear or normal stress on a plane is:
1.   Find the centroid of the plane of interest.
2.   Find the normal and/or shear forces at the centroid of plane of interest.
3.   Convert these forces into a magnitude of the desired direction only, *mag.m.*
4.   Find the area of the cross section.
5.   Divide the force by the area to find the stress.

Although it is not strictly necessary to find the centroid of the cross section of interest for these problems, it is a good idea to do so. In future problems where the bending stresses are also found at a point, it will become necessary, so it is a good habit to get into now. Generally, the centroid is found by inspection in pencil and paper problems anyway.

11

Average Normal and Shear Strains

11.1   Introduction

Until this point all of the materials have been approximated as rigid bodies, meaning no deformation at all. This approximation is not always valid or desired. For this reason the idea of strain has to be developed. The change in length of a member normal strain or the change of the angle between two edges of a member shear strain can be found.

11.2   Problem

Given a simple situation, find the average strain that a part is exposed to. With all of the background tools that have been developed thus far, this should be a matter of putting together pieces of information.

11.3   Theory

The standard assumptions of homogenous, isotropic material will hold in this discussion. The theory is fairly simple for normal stain:

$$\varepsilon_{avg} = \frac{\Delta l}{l}$$

Eq. 11.1

This can be read as: "The average normal strain is the change in length divided by the original length." For this reason, if the part elongates, the normal strain is positive. Then if the part contracts, the normal strain is negative.

The normal strain is often put in the form

$$l' = l(1 + \varepsilon)$$ Eq. 11.2

This can be read as: "The new length of a member is equal to the original length times the quantity one, plus normal strain."

For the shear strain, the definition is

$$\gamma_{nt} = \frac{\pi}{2} - \theta'$$ Eq. 11.3

Where θ' is the new angle between two line segments that were perpendicular before deformation. This formula can be read out as: "The shear strain is equal to one half π radians, 90□°, minus the angle as measured after deformation." By this definition, angles that became acute during deformation underwent a positive shear strain. Those angles that got larger went through a negative shear strain.

These strains are all directional, they are usually broken down into the Cartesian directions for the normal strains, and the angles between the Cartesian directions for the shear stresses.

11.4 Output

Some cases of ways to find the strain in a member follow.

Example A: A plank is supported on one end by a pin connection, and on the other end by a cable see Figure 11.1. Find the strain in the cable when a force is applied to the plank causing a rotation of 0.2□° It is not necessary to use the small angle approximation.

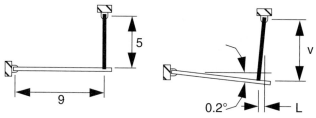

Figure 11.1   Example A.

The first step in this problem is to find the new length of the cable. This can be done with a bit of geometry. The *hyp.m* file should come in handy for this procedure.

CH1101.m

```
v=5+9*sin(DR(0.2));
L=9-9*cos(DR(0.2));
NewLength=hyp(v,L);
NormalStrain=(NewLength-5)/5
```

```
Normal Strain =
   0.0063
```
In pencil and paper calculations, the small angle approximation would probably have been employed. However, with MATLAB it is just as simple to do the calculation precisely so there is little reason to use the approximation.

Example B: In this problem the bottom edge of a plate is held rigid, while the top is sheared to the right see Figure 11.2. If the horizontal edges remain horizontal find the
- Shear strain at corner A
- Shear strain at corner B
- The normal strain along edge AB

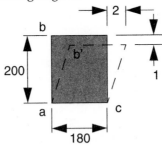

Figure 11.2   Example B.

The new angle at corner a is
```
>>Aangle=atan2((200-1),2);
```
For the angle at corner b a bit more complicated geometry is needed. There are any number of methods to find the angle, the simplest of these is shown here. Knowing that the deformed shape is still a parallelogram, angle c can be calculated.
```
>>Bangle=DR(90)+atan2(2,(200-1));
```
To find the shear stress, simply subtract the new angles from 90°.
```
>>Ashear=DR(90)-Aangle
Ashear =
   0.0100
>>Bshear=DR(90)-Bangle
Bshear =
  -0.0100
```
The final part of this example is similar to Example A.
```
>>NewLength=hyp((200-1),(2));
>>ABstrain=(NewLength-200)/200
ABstrain =
  -0.0049
```

CH1102.m

```
Aangle=atan2((200-1),2);
Bangle=DR(90)+atan2(2,(200-1));
Ashear=DR(90)-Aangle
```

Example C: In this example, the strain of the cable caused by the applied force is known to be 0.012 see Figure 11.3. What is the angle θ that rod *ab* has moved through?

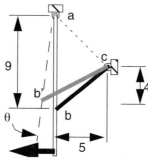

Figure 11.3   Example C.

First the length of the cable after stretching must be found using Equation Set 11.2.

CH1103.m

```
OldLengthBC=hyp(5,4);
NewLengthBC=(1+0.012)*OldLengthBC;
```
To find the new angle θ, triangles *abc,* and *ab'c,* must be solved for.
```
LengthAC=hyp(5,(4-9));
OldAngles=findangle(OldLengthBC,LengthAC,9);
NewAngles=findangle(NewLengthBC,LengthAC,9);
Theta=NewAngles(1)-OldAngles(1)
```

```
Theta =
    0.0109
```
Remember that this θ is in radians, but it can be changed easily enough to degrees:
```
>>RD(Theta)
ans =
    0.6268
```

134

Example D: For this final example what is the shear stress at the corners *a* and *b* after deformation into the parallelogram that is shown in Equation Set 11.4?

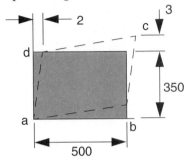

Figure 11.4   Example D.

The new angle of corner *a* can be found by subtracting off the angles *d'ad* and *b'ab* from the original 90°.

CH1104.m

```
dad=atan2(3,500);
bab=atan2(2,350);
NewAngleA=DR(90)-dad-bab;
NewAngleB=DR(180)-NewAngleA;
```
The new angle of corner *b* was found by subtracting the *NewAngleA* that was just found from 180°.

The shear strains at the corners are found by subtracting the deformed angle measurements from the original 90°.
```
ShearA=DR(90)-NewAngleA
ShearB=DR(90)-NewAngleB
ShearA =
   0.0117
ShearB =
  -0.0117
```
Looking at these answers they appear reasonable. The shear at *a* is positive meaning the angle got smaller. The shear at *b* is negative meaning the angle got larger. Then of course, the two shear strains sum to zero as should be expected.

11.5 Summary

These problems are often purely exercises in geometry. Using the routines of *hyp.m* and *findangl.m* creatively will often simplify the geometric difficulties allowing the engineering work to be done more efficiently.

12

Using Variable Input

12.1 Introduction

As can be seen in the last two chapters, many of the problems are similar enough that they can be solved with the same M-files. However, the problems are also unique enough that it is not possible to put a dedicated function file in this text. Using variable input will solve this problem.

12.2 Problem

Knowing that there are several problems to be solved of the same nature, write M-files that can be changed simply so that all of the problems may be solved without rewriting the code each time

12.3 Examples

This technique is best shown by way of examples.

Example A: Here is a problem, and how it would have been solved in the method that has been developed thus far see Equation 12.1.

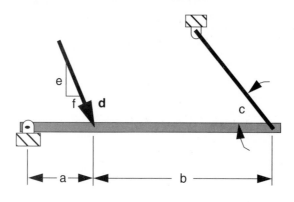

Figure 12.1   Variable input, Example A.

In this problem, what would the magnitude of the force in the supporting cable be if the values shown in Table 12.1were used for the variables?

Data Set	a	b	c	d	e	f	Answer
1	7	3	45°	8	4	3	6.3357
2	6	4	60°	6	4	2	3.7181
3	6	4	30°	6	4	2	6.4399
4	5	5	60°	5	4	4	2.0412

Table 12.1   Four different sets of values for the example variables

Simply write the M-file as normal, except use the lettered variables rather than the actual numbers.

CH1201.m

```
a=7;
b=3;
c=45;
d=8;
e=4;
f=3;

af=rise2xy([-e,f,d,a,0]);
reaction=threevector(af,[0,0,0;DR(90),0,0;DR(180-45),a+b,0]);
CableMagnitude=mag(reaction(3,:))
```

Writing the code in this manner allows for simple changes to any of the variables without searching through the code to make sure all references to the changed variable are accounted for.

To change the variable to the second set of values, simply modify the first lines of code to reflect the new values. Try this now for the remaining sets of values, checking the value versus the given answer.

Even this can get tedious, if there are many different data sets to run through. There may be a more efficient way to enter the data. Instead of editing the data before every run, create a matrix with all of the data already inside it, and then for each run simply change the number of the data set to be used. The Table 12.1 already resembles the data matrix. Code using this method would look like this:

CH1202.m

```
DataSet=1;
data=[7 3 45 8 4 3; 6 4 60 6 4 2; 6 4 30 6 4 2; 5 5 60 5 4 4];
a=data(DataSet,1); %sets a equal to matrix 'data' row
#DataSet column #1
b=data(DataSet,2); %sets b equal to matrix 'data' row
#DataSet column #2
c=data(DataSet,3); %sets c equal to matrix 'data' row
#DataSet column #3
d=data(DataSet,4); %sets d equal to matrix 'data' row
#DataSet column #4
e=data(DataSet,5); %sets e equal to matrix 'data' row
#DataSet column #5
f=data(DataSet,6); %sets f equal to matrix 'data' row
#DataSet column #6
af=rise2xy([-e,f,d,a,0]);
reaction=threevector(af,[0,0,0;DR(90),0,0;DR(180-c),a+b,0]);
CableMagnitude=mag(reaction(3,:))
```

With this new code, when each different data set is to be run, only the number in the *DataSet* variable needs to be changed. If you have much experience in programming, you can see that the next step in this progression would be to loop through the program, changing the data set automatically. Looping techniques will be covered in Chapter 33.

Another advantage to this technique is that if there is a need to go back and rerun a certain data set, it can be done easily. Also, it is possible to look back at past data sets to see if there was an error. With the first method, once a dataset has been changed, there is no going back to it.

Example B: The I-Beam shown in Figure 12.2 is to have constant height and width. The thickness is the only variable, however it is constant throughout the beam. What is the area, I_x, and I_y for the thicknesses of 0.5 through 2 in increments of 0.25?

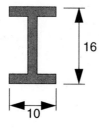

16

10

Figure 12.2   Example B.

CH1203.m

```
DataSet = 1;
ThicknessData=[0.5:0.25:2]; %A matrix from 0.5 to 2 incremented by
0.25
Thickness=ThicknessData(DataSet)% chooses an element from
ThicknessData
Area=ibeam(10,16,Thickness,Thickness,'I','area')
Ix=ibeam(10,16,Thickness,Thickness,'I','Ix')
Iy=ibeam(10,16,Thickness,Thickness,'I','Iy')
```

Now simply run the program changing the dataset each time. A loop might be a good plan here, looping is covered in a later chapter.

13

Plane-Stress Stress Transformations

13.1   Introduction

Knowing the state of stress on an element in a given orientation is useful, but it is also necessary to be able to know the stress state at that same point, but in a different orientation. Different orientations will lead to greater, or lesser shear and normal stresses based on the angle of the new face to be studied. Two planes, called the principle planes, are unique because they do not have any shear stress acting upon them. The stresses acting on these two principle planes are called the principle stresses. These principle stresses are the largest magnitude normal stresses that act on any plane.

The assumptions of plane stress say that the stress in the third direction is equal to zero, however the strain in that third direction is not necessarily zero. Some of the routines do not rely on the plane-stress assumption. If they do, the option of "plane stress" will have to be specifically stated.

13.2   Problem

Given a differential element being acted upon by two normal stresses, σ_x and σ_y, along with a shear stress, τ_{xy}, find:

- Principle stresses, *pristress.m*
- Principle planes, *ppstress.m*
- Stress state on an arbitrary plane, *stresstr.m*

13.3   Theory

The formulas for the principle stresses in plane stress are:

$$\sigma_{p1}, \sigma_{p2} = \frac{(\sigma_x + \sigma_y)}{2} \pm \sqrt{\left(\frac{\sigma_x - \sigma_y}{2}\right)^2 + \tau_{xy}^2}$$

$$\sigma_{p3} = 0$$

Eq. 13.1

The formula for the principle planes:

$$\tan 2\theta_p = \frac{2\tau_{xy}}{\sigma_x - \sigma_y}$$

Eq. 13.2

The stress transformation equations are:

$$\sigma_n = \sigma_x(\cos\theta)^2 + \sigma_y(\sin\theta)^2 + \tau_{xy}\sin\theta\cos\theta$$

$$\tau_{nt} = (\sigma_x - \sigma_y)\sin\theta\cos\theta + \tau_{xy}(\cos\theta)^2 - (\sin\theta)^2$$

Eq. 13.3

NOTE: These equations are by tradition developed for θ, measured counter clockwise from the downward vertical. To remain constant with the notation commonly used in texts, these formulas were used rather than developing the same equations for the notation used in this book. To align the two different conventions, the MATLAB code simply takes 270° from the angle given in our notation to make it correct for the traditional formulas, see Figure 13.1.

Traditional notation: MATLAB notation:

Figure 13.1   Differing notations for plane orientation.

These equations are sufficient to find the principle stresses that act on the principle planes. They also can be used to find the stresses acting on any other plane of interest. In Chapter 16 a visualization tool called Mohr's circle will be developed to show these values graphically.

13.4   Templates

pristress.m

```
function [PrincipleStresses, IPShearMax,
ShearMax]=pristress(State,option)
%PRISTRESS Principal stresses.
%   [PS,IPSM,SM]=PRISTRESS([SIGMAX,SIGMAY,TAUXY],OPTION) is the Principle
%   Stresses, In-Plane Shear Maximum and the Shear Maximum.
%
%   SIGMAX:   Normal stress in the X direction.
%   SIGMAY:   Normal stress in the Y direction.
%   TAUXY:    Shear on the X-Y plane.
%   Together these three are gathered as the STRESSSTATE.
%
%   OPTION: Either 'plane stress' or 'plane strain'.
%
%   See also MOHRS, PPSTRESS, STRESS2STRAIN, STRESSTR.

%   Details are to be found in Mastering Mechanics I, Douglas W. Hull,
%   Prentice Hall, 1999

%   Douglas W. Hull, 1999
%   Copyright (c) 1999 by Prentice Hall
%   Version 1.00

option=lower(option);

sx=State(1);
sy=State(2);
txy=State(3);
center=mean([sx,sy]);
radius=sqrt(((sx-sy)/2)^2+txy^2);
PrincipleStresses(1)=center-radius;
PrincipleStresses(2)=center+radius;
if strcmp(option,'plane stress')
  PrincipleStresses(3)=0;
elseif strcmp(option,'plane strain')
  PrincipleStresses(3)=9999;
  disp ('Not fully supported yet: e-mail author for update; hull@mtu.edu');
else
  disp ('Only options plane strain and plane stress are supported')
  clear PrincipleStresses
  return
end
PrincipleStresses=sort(PrincipleStresses);
IPShearMax=radius;
ShearMax=max([abs(PrincipleStresses/2) radius]);
```

```
>>Stress=[22 30 -20];
>>[ps,ipsm,sm]=pristress(Stress,'plane stress')
ps =
            0      5.6039      46.3961
ipsm =
      20.3961
sm =
      23.1980
```

```
Principle stresses in 'plane stress':
0
5.6039
46.3961

Note: 'plane strain' is the only
other valid choice for the second input.
```

ppstress.m

```
function [PrinciplePlanes]=ppstress(StressState)
%PPSTRESS The principle planes of a stress state.
%    PPSTRESS([SIGMAX,SIGMAY,TAUXY]) Calculates the principle planes of a
%    stress state.
%
%    SIGMAX:   Normal stress in the X direction.
%    SIGMAY:   Normal stress in the Y direction.
%    TAUXY:    Shear on the X-Y plane.
%    Together these three are gathered as the STRESSSTATE.
%
%    See also MOHRS, PRISTRESS, STRESS2STRAIN, STRESSTR.

%    Details are to be found in Mastering Mechanics I, Douglas W. Hull,
%    Prentice Hall, 1999

%    Douglas W. Hull, 1999
%    Copyright (c) 1999 by Prentice Hall
%    Version 1.00

sx=StressState(1);
sy=StressState(2);
txy=StressState(3);
PrinciplePlanes(1,1)=atan2((2*txy/(sx-sy)),1)/2;
PrinciplePlanes(2,1)=PrinciplePlanes(1)+DR(90);
```

144

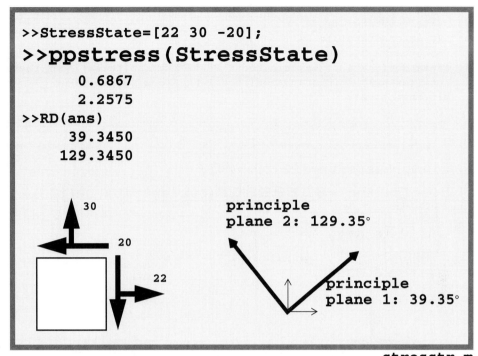

```
>>StressState=[22 30 -20];
>>ppstress(StressState)
       0.6867
       2.2575
>>RD(ans)
      39.3450
     129.3450
```

stresstr.m

```
function [sn,tnt]=stresstr(StressState,theta)
%STRESSTR Stress rotation.
%   [SN, TNT]=STRESSTR([SIGMAX,SIGMAY,TAUXY],THETA) will take the given stress
%   state and rotate it through the given radian angle THETA.
%
%   SIGMAX:  Normal stress in the X direction.
%   SIGMAY:  Normal stress in the Y direction.
%   TAUXY:   Shear on the X-Y plane.
%   Together these three are gathered as the STRESSSTATE.
%   THETA:   The radian angle of rotation.  Positive C.C.W. from right.
%
%   SN:      Normal stress on the given plane.
%   TNT:     Shear stress on the given plane.
%
%   See also MOHRS, PPSTRESS, PRISTRESS, STRESS2STRAIN.


%   Details are to be found in Mastering Mechanics I, Douglas W. Hull,
%   Prentice Hall, 1999


%   Douglas W. Hull, 1999
%   Copyright (c) 1999 by Prentice Hall
%   Version 1.00
```

```
sx=StressState(1);
sy=StressState(2);
txy=StressState(3);
theta=theta-DR(270);
sn=sx*cos(theta).^2+sy*sin(theta).^2+2*txy*sin(theta).*cos(theta);
tnt=-(sx-sy)*sin(theta).*cos(theta)+txy*(cos(theta).^2-
sin(theta).^2);
```

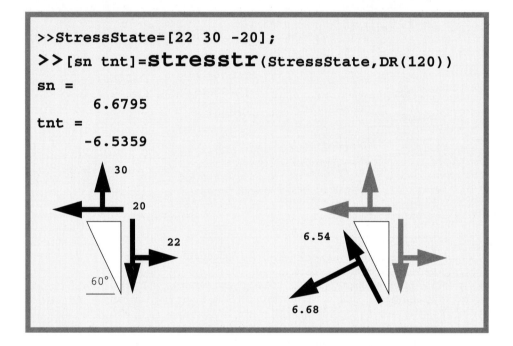

13.5   Output

For the given stress states on differential elements, find the principle stresses, and the principle planes upon which they act see Figure 13.2.

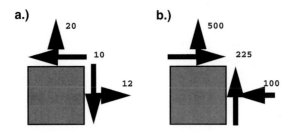

Figure 13.2   Sample stress states.

CH1301.m

```
ssa=[12,20,-10];
ssb=[-100,500,-225];
PrincipleStressesA=pristress(ssa,'plane stress')
PrincipleStressesB=pristress(ssb,'plane stress')
PrinciplePlanesA=RD(ppstress(ssa))
PrinciplePlanesB=RD(ppstress(ssb))
```

```
PrincipleStressesA =
        0    5.2297   26.7703
PrincipleStressesB =
     -175         0       575
PrinciplePlanesA =
   34.0993
  124.0993
PrinciplePlanesB =
   18.4349
  108.4349
```

For the given stress states on differential elements, find the stress state when the element is reoriented as shown in Figure 13.3.

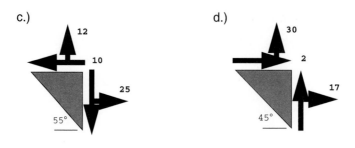

Figure 13.3   More sample stress states.

CH1302.m

```
ssc=[25,12,-10];
ssd=[17,30,2];
[SigmaNormalC, TauNormalC]=stresstr(ssc,DR(180-55))
[SigmaNormalD, TauNormalD]=stresstr(ssd,DR(180-45))
```

```
SigmaNormalC =
   11.3262
TauNormalC =
   -9.5282
SigmaNormalD =
   25.5000
TauNormalD =
    6.5000
```

When the results are returned from the principle stress function, they are automatically arranged in increasing order. When the principle planes are returned, they are also sorted into ascending order. To discover which planes correspond with which stresses, feed the angles back into the stress transformation routine. Looking back at example a:

```
>>PrincipleStressesA
PrincipleStressesA =
        0    5.2297   26.7703
>>PrinciplePlanesA
PrinciplePlanesA =
   34.0993
  124.0993
```

Remember that the *PrinciplePlanesA* is in degrees not radians. This allowed a simpler visualization of the angles. However, MATLAB needs to have it converted back into radians, before using the numbers.

```
>>[SigmaNormalA, TauNormalA]=stresstr(ssa,DR(PrinciplePlanesA))
SigmaNormalA =
    26.7703
     5.2297
TauNormalA =
   1.0e-14 *

   -0.2665
    0.2665
```

This bit of work shows that the same values for the principle stresses are reached through using the *pristress.m* routine, or through taking the principle planes found with *ppstress.m* and seeing what the values for the stresses are on those planes with *stresstr.m*. Using this second method takes more steps, but it can be used to figure out which stresses correspond to which planes. In the above example the 34° plane corresponds to the 26.8 magnitude stress. Because these are the principle stresses, the corresponding shear stresses are zero, as should be expected.

13.6   Features

The *stresstr.m* function can be used to help determine which of the principle stresses coincides with which principle plane. The *stresstr.m* routine can take any number of inputs in the angle argument so long as they are grouped in a matrix. This allows more stresses to be figured with one command.

13.7   Summary

Required argument, optional argument, [**ssv**] = Stress State Vector
pristres(ssv,option)
   Finds the principle stresses of an element that is experiencing the stress state given by the **ssv**. The option is either 'plane stress' or 'plane strain'. The single quotes must be there, and the capitalization must be the same as shown.
ppstress(ssv)
   Finds the principle planes for an element that is experiencing the stress state given by **ssv**. The answer is in radians.
[sn tnt]=stresstr(ssv,angle)
   Finds the σ_n and τ_{nt} on the plane given by the angle. The angle is measured ccw from the right horizontal. The sign notation for σ_n is forces away from the element are positive. For τ_{nt}, use the right hand rule with σ_n to find the positive orientation for the shear stress.
   Stress state vector:

$$\left[\text{normal stress x, normal stress y, shear stress x-y} \right]$$

14

Plane-Stress Strain Transformations

14.1 Introduction

Knowing the state of strain on an element in a given orientation is useful, but it is also necessary to be able to know the strain state at that same point, but in a different orientation. Different orientations will lead to greater, or linear shear and normal strains based on the angle of the new face to be studied. Two planes, called the principle planes, are unique because they do not have any shear strain acting upon them. The normal strains acting on these two principle planes are called the principle strains. These principle strains are the largest magnitude normal strains that act on any plane.

The assumptions of plane strain say that the strain in the third direction is equal to zero, however the strain in that third direction is not necessarily zero. Some of the routines do not rely on the plane-strain assumption. If they do, the option of "plane strain" will have to be specifically stated as an input.

14.2 Problem

Given a differential element being deformed by two normal strains, ε_x and ε_y, along with a shear strain, λ_{xy}, find:

- Principle strains, *pristrain.m*
- Principle planes, *ppstrain.m*
- Strain state on an arbitrary plane, *straintr.m*

14.3   Theory

The formulas for the principle strains in plane strain are:

$$\varepsilon_{p1}, \varepsilon_{p2} = \frac{(\varepsilon_x + \varepsilon_y)}{2} \pm \sqrt{\left(\frac{\varepsilon_x - \varepsilon_y}{2}\right)^2 + \left(\frac{\lambda_{xy}}{2}\right)^2}$$

$$\varepsilon_{p3} = -\frac{\upsilon}{1-\upsilon}(\varepsilon_x + \varepsilon_y)$$

Eq. 14.1

The formula for the principle planes:

$$\tan 2\theta_p = \frac{\lambda_{xy}}{\varepsilon_x - \varepsilon_y}$$

Eq. 14.2

The strain transformation equations are:

$$\varepsilon_n = \varepsilon_x(\cos\theta)^2 + \varepsilon_y(\sin\theta)^2 + \lambda_{xy}\sin\theta\cos\theta$$

$$\lambda_{nt} = (\varepsilon_x - \varepsilon_y)\sin\theta\cos\theta + \lambda_{xy}(\cos\theta)^2 - (\sin\theta)^2$$

Eq. 14.3

NOTE: These equations are by tradition developed for θ, measured counter clockwise from the downward vertical. To remain constant with the notation commonly used in other texts, these formulas were used rather than developing the same equations for the notation used in this book. To align the two different conventions, the MATLAB code simply takes 270° from the angle given in our notation to make it correct for the traditional formulas as shown in Figure 14.1.

Traditional notation: MATLAB notation:

Figure 14.1   Differing notations for plane orientation.

These equations are sufficient to find the principle strains that act on the principle planes. They also can be used to find the strains acting on any other plane of interest. In Chapter 16 a visualization tool called Mohr's circle will be developed to show these values graphically.

14.4   Templates

```
function [PriStrains, IPShearMax,
ShearMax]=pristrai(State,option,poisonts)
%PRISTRAIN Principal strains.
%    [PS,IPSM,SM]=PRISTRESS([EPSX,EPSY,GAMXY],OPTION) is the Principle
%    Stresses, In-Plane Shear Maximum and the Shear Maximum.
%
%    EPSMAX:  Normal strain in the X direction.
%    EPSMAY:  Normal strain in the Y direction.
%    GAMXY:   Shear on the X-Y plane.
%    Together these three are gathered as the STRAINSTATE.
%
%    OPTION:  Either 'plane stress' or 'plane strain'.
%
%    See also MOHRS2, PPSTRAIN, ROSETTE, STRAIN2STRESS, STRAINTR.

%    Prentice Hall, 1999

%    Douglas W. Hull, 1999
%    Copyright (c) 1999 by Prentice Hall
%    Version 1.00

option=lower(option);

if nargin==2
  poisonts=0.3;
end %default poisonts
ex=State(1);
ey=State(2);
gxy=State(3);
center=mean([ex,ey]);
radius=sqrt(((ex-ey)/2)^2+(gxy/2)^2);
PrincipleStrains(1)=center-radius;
PrincipleStrains(2)=center+radius;
if strcmp(option,'plane stress')
  PrincipleStrains(3)=(-1*(poisonts/(1-poisonts)))*(ex+ey);
elseif strcmp(option,'plane strain')
  PrincipleStrains(3)=0;
else
  disp ('Only options plane strain and plane stress are supported')
end
PriStrains=sort(PrincipleStrains);
IPShearMax=radius*2;
ShearMax=PriStrains(3)-PriStrains(1);
```

```
>>StrainState=[22 30 -20];
>>[PS,IPSM,SM]=pristrain(StrainState,'plane strain')
PS =
             0    15.2297    36.7703
IPSM =
       21.5407
SM =
       36.7703
```

```
Principle strains in 'plane strain':
0
15.2297
36.7703

Note: 'plane stress' is the only
other valid choice for the second input.
```

ppstrain.m

```
function [PrinciplePlanes]=ppstrain(StrainState)
%PPSTRAIN The principle planes of a strain state.
%   PPSTRAIN([EPSX,EPSY,GAMXY]) Calculates the principle planes of a
%   strain state.
%
%   EPSMAX:  Normal strain in the X direction.
%   EPSMAY:  Normal strain in the Y direction.
%   GAMXY:   Shear on the X-Y plane.
%   Together these three are gathered as the STRAINSTATE.
%
%   See also MOHRS2, PRISTRAIN,  ROSETTE, STRAIN2STRESS, STRAINTR.

%   Details are to be found in Mastering Mechanics I, Douglas W. Hull,
%   Prentice Hall, 1999

%   Douglas W. Hull, 1999
%   Copyright (c) 1999 by Prentice Hall
%   Version 1.00

ex=StrainState(1);
ey=StrainState(2);
gxy=StrainState(3);
PrinciplePlanes(1,1)=atan2((gxy/(ex-ey)),1)/2;
PrinciplePlanes(2,1)=PrinciplePlanes(1)+DR(90);
```

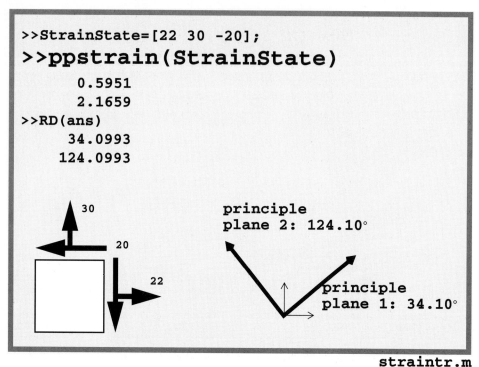

```
>>StrainState=[22 30 -20];
>>ppstrain(StrainState)
      0.5951
      2.1659
>>RD(ans)
     34.0993
    124.0993
```

principle
plane 2: 124.10°

principle
plane 1: 34.10°

straintr.m

```
function [NewState]=straintr(StrainState,theta)
%STRAINTR Stress rotation.
%    [STRAINSTATE]=STRAINTR([EPSX,EPSY,GAMXY],THETA) will take the
given strain
%    state and rotate it through the given radian angle THETA.
%
%    EPSX:    Normal stress in the X direction.
%    EPSY:    Normal stress in the Y direction.
%    GAMXY:   Shear on the X-Y plane.
%    Together these three are gathered as the STRAINSTATE.
%    THETA:   The radian angle of rotation.  Positive C.C.W. from right.
%
%    EXP:     Normal stress in the X prime direction.
%    EYP:     Normal stress in the Y prime direction.
%    GXPYP:   Shear stress in the X prime - Y prime plane.
%    Together these three are gathered as a STRAINSTATE.
%
%    See also MOHRS2, PPSTRAIN, PRISTRAIN, ROSETTE, STRAIN2STRESS.

%    Details are to be found in Mastering Mechanics I, Douglas W. Hull,
%    Prentice Hall, 1999
```

154

```
%    Douglas W. Hull, 1999
%    Copyright (c) 1999 by Prentice Hall
%    Version 1.00

ex=StrainState(1);
ey=StrainState(2);
gxy=StrainState(3);
exp=(ex+ey)/2 + (ex-ey)/2*cos(2*theta) + gxy/2*sin(2*theta);
eyp=(ex+ey)/2 + (ex-ey)/2*cos(2*(theta+DR(90))) + gxy/
2*sin(2*(theta+DR(90)));
gxpyp=-(ex-ey)*sin(2*theta)+gxy*cos(2*theta);
NewState=[exp,eyp,gxpyp];
```

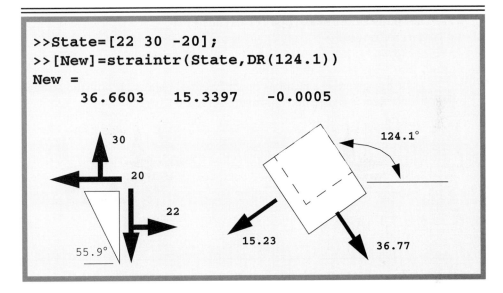

```
>>State=[22 30 -20];
>>[New]=straintr(State,DR(124.1))
New =
     36.6603    15.3397    -0.0005
```

14.5   Output

For the given strain states on differential elements, find the principle strains, and the principle planes upon which they act see Figure 14.2.

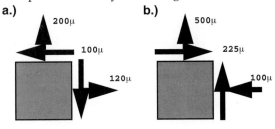

Figure 14.2   Sample strain states.

```
>>ssa=[120,200,-100];
>>ssb=[-100,500,-225];
>>PrincipleStrainsA=pristrain(ssa,'plane stress')
PrincipleStrainsA =
-137.1429  95.9688 224.0312
>>PrincipleStrainsB=pristrain(ssb,'plane stress')
PrincipleStrainsB =
-171.4286-120.4001 520.4001
>>PrinciplePlanesA=RD(ppstrain(ssa))
PrinciplePlanesA =
  25.6701
 115.6701
>>PrinciplePlanesB=RD(ppstrain(ssb))
PrinciplePlanesB =
  10.2780
 100.2780
```

For the given strain states on differential elements, find the strain state when the element is reoriented as shown in Figure 14.3.

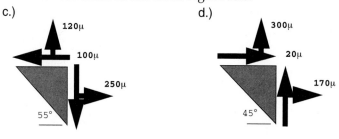

Figure 14.3   More sample strain states.

```
>>ssc=[250,120,-100];
>>ssd=[170,300,20];
>>[NewStrainC]=straintr(ssc,DR(180-55))
NewStrainC =
 209.7533 160.2467 156.3621
>>[NewStrainD]=straintr(ssd,DR(180-45))
NewStrainD =
 225.0000 245.0000-130.0000
```

When the results are returned from the principle strain function, they are automatically arranged in increasing order. When the principle planes are returned, they are also sorted into ascending order. To discover which planes correspond with which strains, feed the angles back into the strain transformation routine. Looking back at example a:

```
>>PrincipleStrainsA
PrincipleStrainsA =
-137.1429  95.9688 224.0312
>>PrinciplePlanesA
PrinciplePlanesA =
   25.6701
  115.6701
```

Remember that the *PrinciplePlanesA* is in degrees not radians. This allowed a simpler visualization of the angles. However, MATLAB needs to have it converted back into radians, before using the numbers.

```
>> [NewStrainState]=straintr(ssa,DR(PrinciplePlanesA))
NewStrainState =
   95.9688 224.0312  -0.0000
  224.0312  95.9688  -0.0000
```

This bit of work shows that the same values for the principle strains are reached through using the *pristrain.m* routine, or through taking the principle planes found with *ppstrain.m* and seeing what the values for the strains are on those planes with *straintr.m*. Using this second method takes more steps, but it can be used to figure out which strains correspond to which planes. In the above example the 25∞ plane corresponds to the 95.96 magnitude normal strain X and a 224 magnitude normal strain Y. Because these are the principle strains, the corresponding shear strains are zero, as should be expected.

14.6   Features

The *straintr.m* function can be used to help determine which of the principle strains coincides with which principle plane. The *straintr.m* routine can take any number of inputs in the angle argument so long as they are grouped in a matrix. This allows more strains to be figured with one command.

14.7   Summary

Required argument, optional argument, [**ssv**] = Strain State Vector

pristres(ssv, option)

> Finds the principle strains of an element that is experiencing the strain state given by the **ssv**. The option is either 'plane strain' or 'plane strain'. The single quotes must be there, and the capitalization must be the same as shown.

ppstrain(ssv)

> Finds the principle planes for an element that is experiencing the strain state given by **ssv**. The answer is in radians.

[ex ey lxy]=straintr(ssv, angle)

> Finds the ε_x, ε_y and λ_{xy} on the plane given by the angle. The angle is measured ccw from the right horizontal. The sign notation for ε_x is forces away from the element are positive. For λ_{xy} use the right-hand rule with ε_x or ε_y to find the positive orientation for the shear strain.

Strain state vector:

$$\begin{bmatrix} \text{normal strain x,} & \text{normal strain y,} & \text{shear strain x-y} \end{bmatrix}$$

15

Strain Rosettes

15.1   Introduction

There is no practical way to measure stress directly. However, strain can be measured fairly accurately by way of a strain gauge. More often than not, a set of three strain gauges is used at once to measure the strain in three different directions. These three directions can later be used to calculate the strain state completely. Usually when the strain state is found with strain gauges on the surface the assumption of plane stress can be made.

15.2   Problem

Given a strain rosette in an arbitrary orientation and spacing of the rosettes in the strain rosette, find the strain state. This strain state can then be used as an input into other strain routines.

15.3   Theory

The normal strain transformation equation:

$$\varepsilon_n = \varepsilon_x(\cos\theta)^2 + \varepsilon_y(\sin\theta)^2 + \lambda_{xy}\sin\theta\cos\theta \qquad \text{Eq. 15.1}$$

This equation can be applied once to each of the three strain gauges. This will make a set of three equations, three unknowns see Figure 15.1 . In matrix form these equations appear as:

$$
\begin{bmatrix}
(\cos\theta_a)^2 & (\sin\theta_a)^2 & \sin\theta_a\cos\theta_a \\
(\cos\theta_b)^2 & (\sin\theta_b)^2 & \sin\theta_b\cos\theta_b \\
(\cos\theta_c)^2 & (\sin\theta_c)^2 & \sin\theta_c\cos\theta_c
\end{bmatrix}
\begin{bmatrix}
\varepsilon_x \\
\varepsilon_y \\
\lambda_{xy}
\end{bmatrix}
=
\begin{bmatrix}
\varepsilon_a \\
\varepsilon_b \\
\varepsilon_c
\end{bmatrix}
\qquad \text{Eq. 15.2}
$$

This set of equations can easily be solved for the strain state.

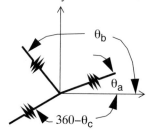

Figure 15.1   Notation for the strain gauges.

This allows any combination of strain gauges to be used, regardless of the orientation of the gauges.

15.4   Template

rosette.m

```
function [StrainState]=rosette(epsilon, theta)
%ROSETTE Converts strain gauge readings to strain state.
%    ROSETTE([EPS1,EPS2,EPS3],[THETA1,THETA2,THETA3]) converts
strain gauge
%    readings to strain state.
%
%    See also MOHRS2, PPSTRAIN, PRISTRAIN, STRAIN2STRESS, STRAINTR.

%    Details are to be found in Mastering Mechanics I, Douglas W. Hull,
%    Prentice Hall, 1999

%    Douglas W. Hull, 1999
%    Copyright (c) 1999 by Prentice Hall
%    Version 1.00

coef=[cos(theta').^2 sin(theta').^2 sin(theta').*cos(theta')];
StrainState=(inv(coef)*epsilon')';
```

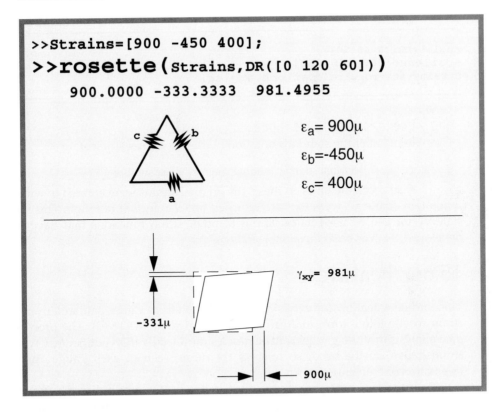

```
>>Strains=[900 -450 400];
>>rosette(Strains,DR([0 120 60]))
   900.0000 -333.3333   981.4955
```

$\varepsilon_a= 900\mu$

$\varepsilon_b=-450\mu$

$\varepsilon_c= 400\mu$

$\gamma_{xy}= 981\mu$

-331μ

900μ

15.5   Output

For the given outputs from the strain rosette, find the strain state of the object at that point see Figure 15.2.

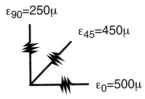

$\varepsilon_{90}=250\mu$

$\varepsilon_{45}=450\mu$

$\varepsilon_0=500\mu$

Figure 15.2   Strain gauge with sample readings.

To solve for the strain state at this point, first you must gather the strains in one vector, and then the angles in another:

```
angles=DR([0 45 90]);
epsilons=[500 450 250];
StrainState=rosette(epsilons,angles)
```

```
StrainState =
 500.0000 250.0000 150.0000
```

Remember the standard notation in this book for a strain state is:

$$\left[\text{normal strain x, normal strain y, shear strain x-y}\right]$$

Also check the units on these, the original strains were entered in with the μ notation so the answers that are returned are also in that notation. This strain state vector can be used as the input to other strain functions that have been developed, such as *pristrain.m* or *ppstrain.m*.

15.6 Features

This routine will allow for very simple conversion from values gathered from any strain rosette into a form that can be manipulated by MATLAB. The process of data analysis in strain gauges will be much simplified by this routine. Any type of strain gauge can be used, so long as the strain seen in each gauge and the orientation of that gauge are known.

If the strain gauges return only a voltage, not a strain directly, afunction could be created to do the voltage-to-strain calculation and automatically feed the answer into the *rosette.m* code. The variety of different strain gauges makes it unfeasible to write the code that would likely be needed for that calculation. Study of MATLAB help files and the functions presented in this test will aid the creation of these conversion files.

15.7 Summary

Required argument, optional argument, [ssv] = Strain State Vector
[ssv]=rosette(strains,angles)
Finds the strain state at the point of application of the strain rosette. The rosette contains three strain gauges. The inputs to the function are vectors of three values:

$$\text{strains} = \left[\varepsilon_1\ \varepsilon_2\ \varepsilon_3\right]$$

$$\text{angles} = \left[\theta_1\ \theta_2\ \theta_3\right]$$

Strain state vector:

$$\left[\text{normal strain x, normal strain y, shear strain x-y}\right]$$

16

Mohr's Circle for Stress

16.1 Introduction

Mohr's circle is a powerful geometric representation of stress or strain on an object. In paper and pencil calculations, Mohr's circle was useful because it transformed complicated formulas into simple geometry problems. A second point about the Mohr's circle that still makes it useful today is the visual representation of many stresses on all the different planes of orientation. At a glance, the principle stresses and the maximum in-plane shear stress may be obtained.

16.2 Problem

Knowing the stress state of an object at a point, draw the Mohr's circle representation of the stress. Include in the diagram a list of
- The center of the in-plane circle
- The maximum in-plane shear
- The maximum shear in any plane
- The three principle stresses
- The two principle planes

The routine should accept both the assumption of "plane stress" or "plane strain." Furthermore, if a certain plane is of interest, that plane may be input so that it is highlighted on the diagram and the values of the shear and normal stress are explicitly listed on the diagram also.

16.3 Theory

The Mohr's circle is really a graph of the shear and normal stresses for all different angles of orientation of the plane of interest. A picture of a Mohr's circle is it's own best explanation see Figure 16.1.

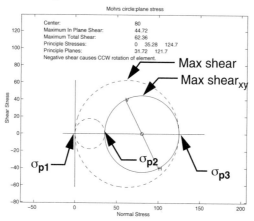

Figure 16.1   Mohr's circle for stress state of σ_x=60, σ_y=100, τ_{xy}=-40.

These drawings are easy to understand. The x axis represents the normal stresses while the shear stresses are on the y axis. The solid circle represents the stress states on the x-y plane. The dashed circles represent the x-y stress states and the y-z stress states. The three points where the circles cross the x axis are the principle stresses. The maximum shear stress on any of the three planes is the maximum total shear, and is found by reading the shear value of the topmost point of the largest circle. The maximum shear stress on the x-y plane is found by looking at the topmost point of the solid circle.

The radius from the center of the solid circle labeled H represents the horizontal plane. The other radius labeled with a V represents the vertical plane. Notice that 180° separate the two planes, not 90°. Similarly, if a plane is 30° rotated ccw from horizontal, the corresponding radius on the Mohr's circle is found at 60° ccw from the radius designated as horizontal. This can be seen when a plane is specified in the function call.

The data that is generated for this graph is calculated using the *pristress.m* and *ppstress.m* functions.

16.4   Template

```
function []=mohrs(StressState, option, angle)
%MOHRS Draws a Mohr's circle.
%    MOHRS([SIGMAX,SIGMAY,TAUXY],OPTION,ANGLE) Calculates principle stresses,
%    MAXIMUM SHEAR, AND NORMAL AND SHEAR STRESS ON A REQUESTED PLANE. ALL OF
%    these are presented graphically on a Mohr's circle diagram that can be
%    easily printed out.
%
%    SIGMAX:  Normal stress in the X direction.
%    SIGMAY:  Normal stress in the Y direction.
%    TAUXY:   Shear on the X-Y plane.
%    Together these three are gathered as the STRESSSTATE.
%
%    OPTION: Either 'plane stress' or 'plane strain'.
%    ANGLE: Optional angle of interest will be highlighted on figure.
%
%    See also MOHRS2, PPSTRESS, PRISTRESS, STRESS2STRAIN, STRESSTR.

%    Details are to be found in Mastering Mechanics I, Douglas W. Hull,
%    Prentice Hall, 1999

%    Douglas W. Hull, 1999
%    Copyright (c) 1999 by Prentice Hall
%    Version 1.00

sx=StressState(1);
sy=StressState(2);
txy=StressState(3);
center=mean([sx,sy]);
[PrincipleStresses, IPShearMax, ShearMax]=pristress(StressState,
option);
PP=ppstress(StressState)';
radius=IPShearMax;
clf
showcirc(radius,[center,0],'r');
hold on
showcirc(PrincipleStresses(1)/2,[PrincipleStresses(1)/2,0],'r--');
showcirc(PrincipleStresses(2)/2,[PrincipleStresses(2)/2,0],'r--');
showcirc(PrincipleStresses(3)/2,[PrincipleStresses(3)/2,0],'r--');
axis ('equal')
edges=axis;
le=edges(1);
hs=(edges(2)-edges(1))/2;
plot ([edges(1)-0.1*edges(1) edges(2)+0.1*edges(2)],[0 0],'b')
plot ([0,0],[edges(3) edges(4)],'b')
plot (center,0,'ro')
plot ([sx,sy],[-txy,txy],'k')
colA=strvcat('Center:','Maximum In Plane Shear:','Maximum Total
Shear:');
colA=strvcat(colA,'Principle Stresses:','Principle Planes:');
colB=strvcat(num2str(center,4),num2str(IPShearMax,4),num2str(ShearMax,
4));
colB=strvcat(colB,num2str(PrincipleStresses,4),num2str(RD(PP),4));
if nargin==3
```

165

```
   AngleToHorPlane=atan2(txy,(sy-center));
   AngleToRequestPlane=AngleToHorPlane + 2*angle;
   rn=center + radius * cos(AngleToRequestPlane);
   rs=radius * sin(AngleToRequestPlane);
   plot ([center,rn],[0,rs],'r',rn,rs,'rd')
   colA=strvcat(colA,'At angle:','**Normal Stress:','**Shear Stress:');
   colB=strvcat(colB,num2str(RD(angle),4),num2str(rn,4),num2str(rs,4));
end
axis ('equal')
colA=strvcat(colA,'Negative shear causes CCW rotation of element.');
colB=strvcat(colB,' ');
expandaxis (30, 30)
titleblock(colA,colB);
xlabel ('Normal Stress')
ylabel ('Shear Stress')
title (strcat('Mohrs circle:   ',option))
text (sx,-txy,'V')
text (sy,txy,'H')
hold off
```

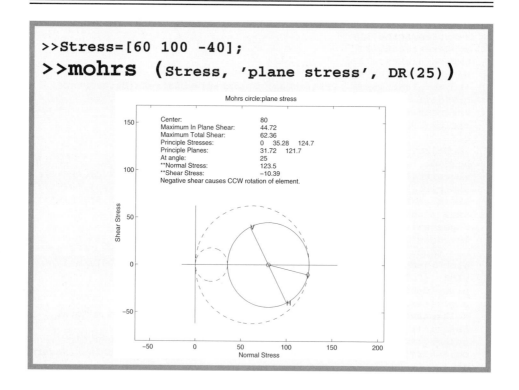

166

16.5   Output

Generate the Mohr's circle representation of the stress state. In particular represent the plane of interest on the diagram. Assume plane stress. See Figures 16.2 and 16.3.

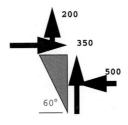

Figure 16.2   Stress state on an element.

```
>>StressState=[-500 200 350];
>>mohrs (StressState,'plane stress',DR(60))
```

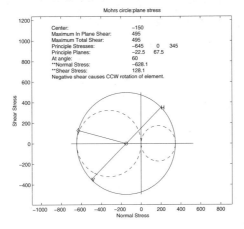

Figure 16.3   Mohr's Circle for stress state of σ_x=-500 σ_y=200 τ_{xy}=350.

16.6   Features

For anyone that has drawn many Mohr's circles by hand, the ease with which they may be generated with this routine can be truly appreciated. The important values being extracted and put into a title block also is very useful. These diagrams should be ready to print out and hand in for students. If more data should appear on the diagram, the *gtext* or *text* commands should prove useful.

16.7   Summary

Required argument, optional argument, [**ssv**] = Stress State Vector
mohrs(*ssv,option,*angle)
>   This routine will draw a Mohr's circle representation of the stress state given to it. It works with one of two options: 'plane stress' or 'plane strain'. The single quotes must be there, and the capitalization must be the same as shown. An optional angle of interest can be passed to the function also.

Stress state vector:

$$\begin{bmatrix} \text{normal stress x} & \text{normal stress y} & \text{shear stress x-y} \end{bmatrix}$$

17

Mohr's Circle for Strain

17.1 Introduction

Mohr's circle is a powerful geometric representation of stress or strain on an object. In paper and pencil calculations, Mohr's circle was useful because it transformed complicated formulas into simple geometry problems. A second point about the Mohr's circle that still makes it useful today is the visual representation of the strain on all the different planes of orientation. At a glance the principle strains and the maximum in-plane shear strain may be obtained.

17.2 Problem

Knowing the strain state of an object at a point, draw the Mohr's circle representation of the strain. Include in the diagram a list of
  - The center of the in-plane circle
  - The maximum in-plane shear
  - The maximum shear in any plane
  - The three principle strains
  - The two principle planes

The routine should accept both the assumption of "plane stress" or "plane strain." Furthermore, if a certain plane is of interest, that plane may be input so that it is highlighted on the diagram and the values of the shear and normal strain are explicitly listed on the diagram also.

17.3   Theory

The Mohr's circle is really a graph of the shear and normal strains for all different angles of orientation of the plane of interest. A picture of a Mohr's circle is its own best explanation see Figure 17.1.

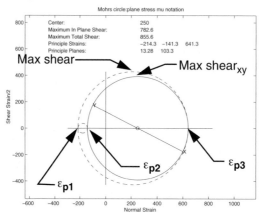

Figure 17.1   Mohr's circle for strain state of $\varepsilon_x=600\mu$, $\varepsilon_y=-100\mu$, $\lambda_{xy}=350\mu$, $\upsilon=0.3$.

These drawings are easy to understand. The x axis represents the normal strains while the shear strains must be divided by two before being put on the y axis. This scaling on the y axis allows the graph to appear as a circle rather than an elliptical form. The solid circle represents the strain states on the x-y plane. The dashed circles represent the x-z strain states and the y-z strain states. The three points where the circles cross the x-axis are the principle strains. The maximum shear strain on any of the three planes is the maximum total shear, and is found by reading the shear value of the topmost point of the largest circle. The maximum shear strain on the x-y plane is found by looking at the topmost point of the solid circle.

The radius from the center of the solid circle labeled X represents the horizontal plane. The other radius labeled with a Y represents the vertical plane. Notice that 180° separate the two planes, not 90°. Similarly, if a plane is 30° rotated ccw from horizontal, the corresponding radius on the Mohr's circle is found at 60° ccw from the radius designated with the X. This can be seen when a plane is specified in the function call.

The data that is generated for this graph is calculated using the *pristrain.m* and *ppstrain.m* functions.

17.4 Template

```
function []=mohrs2(StrainState, option, poissons, angle)
%MOHRS2 Draws a Mohr's circle.
%   MOHRS2([EPSX,EPSY,GAMXY],OPTION,POISSONS,ANGLE) Calculates principle
%   strains, maximum shear, and normal and shear strains on a requested
%   plane.  All of these are presented graphically on a Mohr's circle
%   diagram that can be easily printed out.
%
%   EPSX:    Normal strain in the X direction.
%   EPSY:    Normal strain in the Y direction.
%   GAMXY:   Shear on the X-Y plane.
%   Together these three are gathered as the STRAINSTATE.
%
%   OPTION:  Either 'plane stress' or 'plane strain'.
%   ANGLE:   Optional angle of interest will be highlighted on figure.
%
%   See also MOHRS, PPSTRAIN, PRISTRAIN,  ROSETTE, STRAIN2STRESS,
%      STRAINTR.

%   Details are to be found in Mastering Mechanics I, Douglas W. Hull,
%   Prentice Hall, 1999

%   Douglas W. Hull, 1999
%   Copyright (c) 1999 by Prentice Hall
%   Version 1.00

StrainState=StrainState*1e6;
ex=StrainState(1);
ey=StrainState(2);
gxy=StrainState(3);
center=mean([ex,ey]);
[PS, IPShearMax, ShearMax]=pristrain(StrainState,option,poissons);
PP=ppstrain(StrainState)';
radius=IPShearMax/2;
clf
showcirc(radius,[center,0],'r');
hold on
showcirc(PS(2)-mean([PS(1),PS(2)]),[mean([PS(2),PS(1)]),0],'r--');
showcirc(PS(3)-mean([PS(1),PS(3)]),[mean([PS(3),PS(1)]),0],'r--');
showcirc(PS(3)-mean([PS(2),PS(3)]),[mean([PS(3),PS(2)]),0],'r--');
axis ('equal')
edges=axis;
le=edges(1);
hs=(edges(2)-edges(1))/2;
plot ([edges(1)-0.1*edges(1) edges(2)+0.1*edges(2)],[0 0], 'b')
plot ([0,0],[edges(3) edges(4)],'b')
plot (center,0,'ro')
plot ([ex,ey],[-gxy/2,gxy/2],'k')
colA=strvcat('Center:','Maximum In Plane Shear:','Maximum Total
Shear:');
colA=strvcat(colA,'Principle Strains:','Principle Planes:');
colB=strvcat(num2str(center,4),num2str(IPShearMax,4),num2str(ShearMax,
4));
colB=strvcat(colB,num2str(PS,4),num2str(RD(PP),4));
if nargin==4
```

```
    AngleToHorPlane=atan2(gxy/2,(ey-center));
    AngleToRequestPlane=AngleToHorPlane + 2*angle;
    rn=center + radius * cos(AngleToRequestPlane);
    rs=radius * sin(AngleToRequestPlane);
    plot ([center,rn],[0,rs],'r',rn,rs,'rd')
    colA=strvcat(colA,'At angle:','**Normal Strain:','**Shear Strain:');
    colB=strvcat(colB,num2str(RD(angle),4),num2str(rn,4),num2str(rs,4));
end
axis ('equal')
expandaxis (30, 30)
titleblock(colA,colB);
xlabel ('Normal Strain')
ylabel ('Shear Strain/2')
title (strcat('Mohrs circle:    ',option,' mu notation'))
text (ex,-gxy/2,'X')
text (ey,gxy/2,'Y')
hold off
```

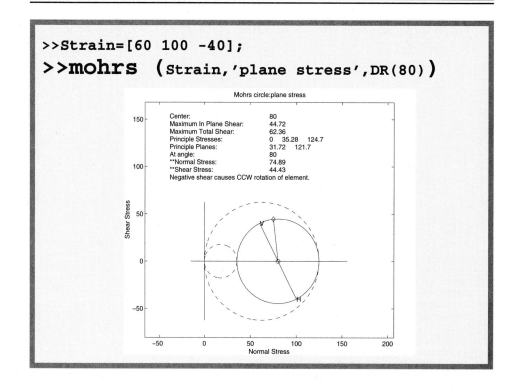

172

17.5 Output

Generate the Mohr's circle representation of this strain state. In particular represent the plane of interest on the diagram, Poisson's ratio is 0.3. Assume plane stress. See Figures 17.2 and 17.3.

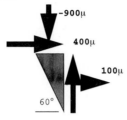

Figure 17.2   Strain state on an element.

```
>>StrainState=[100 -900 400]*1e-6;
>>mohrs2 (StrainState,'plane stress',0.3, DR(120))
```

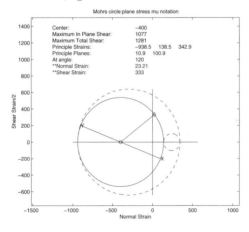

Figure 17.3   Mohr's Circle for stress state of ε_x=-100μ, ε_y=-900μ, λ_{xy}=400μ.

17.6 Features

For anyone that has drawn many Mohr's circles by hand, the ease and accuracy with which they may be generated with this routine can be truly appreciated. The important values being extracted and put into a legend of sorts also is very useful.

These diagrams should be ready to print out and hand in for students. If more data should appear on the diagram, the *gtext* or *text* commands should prove useful.

17.7   Summary

Required argument, optional argument, [**ssv**] = Strain State Vector
mohrs2(*ssv,option,*angle)

> This routine will draw a Mohr's circle representation of the strain state given to it. It works with one of two options: 'plane stress' or 'plane strain'. The single quotes must be there, and the capitalization must be the same as shown. An optional angle of interest can be passed to the function also.

Strain state vector:

$$\left[\text{normal strain x, normal strain y, shear strain x-y}\right]$$

18

Strain—Stress Conversions

18.1 Introduction

Stress and strain are related to each other by the material properties of the object being studied. For this book, the assumption of homogenous isotropic materials will be made. This allows for the derivation of an elegant relationship between stress and strain.

18.2 Problem

Given the stress or strain state on a body, find the other state. The material properties of Young's modulus, E, and Poisson's ratio, υ, will be provided.

18.3 Theory

The conversions from stress to strain are:

$$\varepsilon_x = \frac{1}{E}(\sigma_x - \nu\sigma_y)$$

$$\varepsilon_y = \frac{1}{E}(\sigma_y - \nu\sigma_x)$$ Eq. 18.1

$$\gamma_{xy} = \frac{\tau_{xy}}{E}2(1+\upsilon)$$

The conversions from strain to stress are:

$$\sigma_x = \frac{E}{(1-\upsilon^2)}(\varepsilon_x + \upsilon\varepsilon_y)$$

$$\sigma_y = \frac{E}{(1-\upsilon^2)}(\varepsilon_y + \upsilon\varepsilon_x)$$

Eq. 18.2

$$\tau_{xy} = \frac{\gamma_{xy}E}{2(1+\upsilon)}$$

These sets of equations are easily solved for the desired values.

18.4   Template

strain2stress.m

```
function [StressState] = rain2ess(StrainState,E,poisonts)
%STRAIN2STRESS Converts strain to stress.
%   STRAIN2STRESS([EPSX,EPSY,GAMXY],E,POISSONS) Converts a strain state
%   to the equivalent stress state.
%
%   EPSX:  Normal strain in the X direction.
%   EPSY:  Normal strain in the Y direction.
%   GAMXY:   Shear on the X-Y plane.
%   Together these three are gathered as the STRAINSTATE.
%
%   E:       Young's Modulus
%   POISSONS:Poisson's ratio.
%
%   See also MOHRS2, PPSTRAIN, PRISTRAIN, ROSETTE, STRAINTR,
%      STRESS2STRAIN.

%   Details are to be found in Mastering Mechanics I, Douglas W. Hull,
%   Prentice Hall, 1999


%   Douglas W. Hull, 1999
%   Copyright (c) 1999 by Prentice Hall
%   Version 1.00


ex=StrainState(1);
ey=StrainState(2);
lxy=StrainState(3);

StressState(1)=(E/(1-poisonts^2))*(ex+poisonts*ey);
StressState(2)=(E/(1-poisonts^2))*(ey+poisonts*ex);
StressState(3)=lxy*E/(2*(1+poisonts));
```

```
>>Strain=[220 300 -200]*1e-6;
>>E=210e6; v=0.3;
>>Stress=strain2stress(Strain,E,v)
Stress =
    1.0e+04 *
       7.1538      8.4462     -1.6154
```

```
          300μ                    84.5 K
          200μ                    -16.2 K
          220μ                    71.5 K

      Strain State            Stress State
```

stress2strain.m

```
function [StrainState] = stress2strain(StressState,E,poisonts)
%STRESS2STRAIN Converts stress to strain.
%   STRESS2STRAIN([SIGMAX,SIGMAY,TAUXY],E,POISSONS) Converts a stress state
%   to the equivalent strain state.
%
%   SIGMAX:  Normal stress in the X direction.
%   SIGMAY:  Normal stress in the Y direction.
%   TAUXY:   Shear on the X-Y plane.
%   Together these three are gathered as the STRESSSTATE.
%
%   E:        Young's Modulus
%   POISSONS:Poisson's ratio.
%
%   See also MOHRS, PPSTRESS, PRISTRESS, STRAIN2STRESS, STRESSTR.

%   Details are to be found in Mastering Mechanics I, Douglas W. Hull,
%   Prentice Hall, 1999

%   Douglas W. Hull, 1999
%   Copyright (c) 1999 by Prentice Hall
%   Version 1.00


sx=StressState(1);
sy=StressState(2);
txy=StressState(3);

StrainState(1)=(sx-poisonts*sy)/E;
StrainState(2)=(sy-poisonts*sx)/E;
StrainState(3)=(txy/E)*2*(1+poisonts);
```

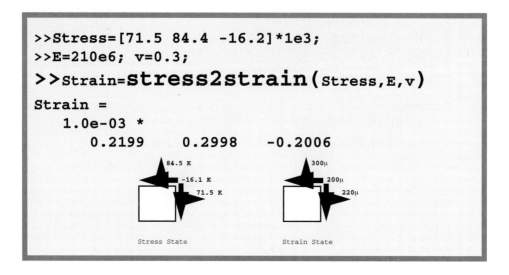

```
>>Stress=[71.5 84.4 -16.2]*1e3;
>>E=210e6; v=0.3;
>>Strain=stress2strain(Stress,E,v)
Strain =
    1.0e-03 *
        0.2199        0.2998        -0.2006
```

18.5   Output

Find the stress state if the strain state and material properties are:

$$\varepsilon_x = 300\mu \qquad \varepsilon_y = 0\mu \qquad \gamma_{xy} = 0\mu$$

$$E = 210 \times 10^6 \qquad \upsilon = 0.3$$

```
>>StrainState=[300 0 0]*1e-6;
>>E=210e6;
>>v=0.3;
>>StressState=strain2stress(StrainState,E,v)
StressState =
1.0e+04 *

    6.9231  2.0769          0
```

This answer makes sense. If a bar of material is lengthening in only one direction, then a stress must be applied perpendicular to the original stress to counteract the contraction. This, as is expected, has occurred.

As a further example, reverse the calculation to ensure that the calculated stress state will return the original strain state.

```
>>StrainState2=stress2strain (StressState,E,v)
StrainState2 =
1.0e-03 *

    0.3000          0          0
```

This shows that the routines give compatible answers.

18.6   Features

These two routines give the link between the worlds of stress and strain. With these routines developed so far in this book, it would be possible to take the readings from a strain rosette, convert them to a stress state, then to draw this stress state as a Mohr's circle.

18.7   Summary

Required argument, optional argument, [ssv] = Stress/Strain State Vector

[ssv]=strain2stress(ssv,Young's Modulus, Poisson's ratio)
   Converts a Strain State Vector to it's equivalent Stress State Vector.

[ssv]=stress2strain(ssv,Young's Modulus, Poisson's ratio)
   Converts a Stress State Vector to it's equivalent Strain State Vector.

Stress state vector:

$$\left[\text{normal stress x, normal stress y, shear stress x-y}\right]$$

Strain state vector:

$$\left[\text{normal strain x, normal strain y, shear strain x-y}\right]$$

19

Material Properties

19.1   Introduction

Mechanical elements act in a manner that is characteristic of their component material. Many materials have been extensively studied so that their properties may be cataloged and used for calculations without further testing. Some of the more commonly used material constants are Young's modulus, E, Poisson's ratio, υ, and yield strength, σ_y.

19.2   Problem

This chapter's template is a look-up routine to find the material constants for several common materials. The possible constants are:
- Young's modulus, E, (GPa, ksi)
- Modulus of rigidity, G, (GPa, ksi)
- Yield strength, σ_y, (MPa, ksi)
- Ultimate strength, σ_u, (MPa, ksi)
- Poisson's ratio, υ, (unitless)
- Coefficient of thermal expansion, α, (1/°C, 1/°F)

The materials available are:
- Aluminum (2014-T6)
- Cast iron (Gray ASTM 20)
- Bronze (C86100)
- Structural steel (A36)
- Stainless steel (304)
- Tool steel (L2)
- Titanium (Ti-6Al-4V)

The units can be in either U.S. Customary or SI. This routine exists for simplicity only. When working problems it's nice to have a sampling of material properties without having to find a reference book. For a more comprehensive set of values, reference books should be consulted.

19.3   Template

```
function [value]=matprop (name,constant,USSI)
%MATPROP Material properties look up.
%    MATPROP(NAME,PROPERTY,UNITS) will find the value of a material property
%    given the material name, the property and the unit system, either US or
%    SI units.  Type 'matprop ('list')' to see available materials and
%    properties.

%    Details are to be found in Mastering Mechanics I, Douglas W. Hull,
%    Prentice Hall, 1999

%    Douglas W. Hull, 1999
%    Copyright (c) 1999 by Prentice Hall
%    Version 1.00

if nargin<2
  name='list';
end
if nargin==2
  disp('US or SI units must be specified.')
  return
end
load matprop.mat

[NumMaterials NumConstants]=size(ValuesMatrix);
ColSp=blanks(NumMaterials)';
ColTab=[ColSp ColSp ColSp ColSp ColSp];
RowSp=blanks(NumConstants)';
RowTab=[RowSp RowSp RowSp RowSp RowSp];

if strcmp(name,'list')
  disp(' ');
  disp('Known Materials');
  disp([ColTab(1:NumMaterials/2,:) NamesVector(1:NumMaterials/2,:)]);
  disp(' ');
  disp('Known Constants and their Units (SI/US)');
  disp([RowTab ConstantsVector RowSp UnitsVectorSI RowSp UnitsVectorUS]);
  disp(' ');
```

```
disp('The third argument must be either ''SI'' or ''US'' to choose units.');
  disp(' ');
  disp('When specifying arguments to the function, single quotes must');
  disp('be used around all inputs and capitalization must be as shown');
  return
end

if strcmp(USSI,'US')
  SI=0;
elseif strcmp(USSI,'SI')
  SI=1;
else
  disp ('Invalid unit system specification. Try ''SI'' or ''US''.' )
  return
end

rowname=strmatch(name,NamesVector,'exact');
rowSI=find(SIVector==SI);
row=intersect(rowSI,rowname);
col=strmatch(constant,ConstantsVector,'exact');
if isempty(row)
    disp ('Material not found. Try ''list'' for valid choices')
    return
end

if isempty(col)
    disp ('Property not found. Try ''list'' for valid choices')
    return
end

value=ValuesMatrix(row,col);
```

>>**matprop**('bronze', 'yield tension','SI')

345

>>**matprop**('list')

This command will list known materials, known constants and their units along with some general help.

19.4 Problem

For a strain problem, the SI unit's version of Young's modulus for structural steel needs to be entered into a function. Find this value and assign it to the variable *v*.

```
>>v=matprop('structural steel','E','SI')
v =
    200
```

If a reminder is needed for what units this answer is in, or for any other help, the command to execute is either of the following:

```
>>matprop('list')
>>matprop
```

Both of these commands have the same effect.

Some material constants are not defined for any of a number of reasons. If this is the case, such as for yield strength in tension of cast iron in either unit system:

```
>>matprop('cast iron','yield tension','US')
    NaN
```

The NaN stands for Not a Number. This will alert the user that there is no valid value for the request.

19.5 Features

This routine allows the user to get material constants for several common engineering materials. This is nice when quick calculations are being made and it would be distracting to find a proper reference manual for the exact type of steel being used. It is certainly no replacement for a complete reference book, but is often times more convenient.

19.6 Summary

Required argument

matprop(*material name*,*material constant*,*units*)

This routine will read a data set to find the requested material property. All three arguments must be in single quotes, and they must be capitalized as shown when a list is requested. To request a list of materials and constants have the only argument to the function be 'list', or just run the routine with no arguments at all.

20

Stress Strain—Examples

20.1   Standard shapes

For each of the illustrated shapes in Figure 20.1, find the requested information.

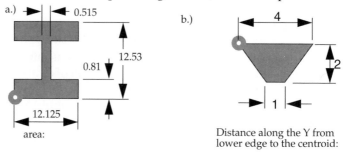

Figure 20.1   Standard shapes.

Shape *a* is from a structural shapes handbook, it is wide-flange sectionW12x87. The dimensions given are from the tables. The area given in the handbook is 25.6. Compare that to the calculated values

```
>>areaA = ibeam(12.125,12.53,0.81,0.515,'I','area')
areaA =
  25.2612
```

Our answer is slightly smaller than the value given in the handbook. This is acceptable because our model is based on a purely rectangular design, where the real beams are tapered and filleted. When greater accuracy is needed, it is best to consult hand books or actual measurements. Although in most cases these approximations are more than sufficient.

The second problem is a bit tricky:

```
>>distB=hortrap(4,-2,1,1.5,'centY')
distB =
  -0.8000
```

The problem is not over at this point. This is the distance from the datum of the shape. The question asked for the distance from the lower edge.

```
>>distB=2+distB
distB =
   1.2000
```

This is the value that was requested. It is important to make sure any additional calculations that need to be done for a problem are actually done, do not just take the answer that comes from the functions without further thought.

20.2   Composite shapes

For each of the illustrated shapes in Figure 20.2, find the requested information.

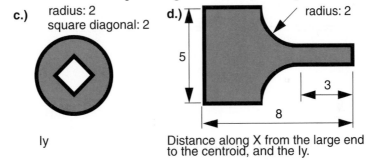

c.)   radius: 2
square diagonal: 2

Iy

d.)   radius: 2

5

3

8

Distance along X from the large end to the centroid, and the Iy.

Figure 20.2   Composite shapes.

CH2001.m

```
partC(1,:)=[circle(2,'comp'),0,0,1];
partC(2,:)=[hortria(2,1,1,'comp'),-1,0,-1];
partC(3,:)=[hortria(2,-1,1,'comp'),-1,0,-1];
IyC=comp(partC,'Iy')
```

```
IyC =
   12.2330
```

These lines of code may need some additional explanation. The datum point was chosen to be the center of the circle. It was then decided that there should be three primitives used: a solid circle, and then two horizontal triangles to create the hole.

```
>>partC(1,:)=[circle(2,'comp'),0,0,1];
```

This starts the parts matrix, as explained in "Creating the Parts Matrix" Section 9.5.

```
>>partC(2,:)=[hortria(2,1,1,'comp'),-1,0,-1];
```

This creates the upper triangle that forms the hole. Notice that the datum of the triangle is to the left of the datum of the completed part. Also notice the final argument is '-1' indicating that the shape represents a hole.

Example *d* does not appear to lend itself to this method. But on further examination it can be broken down into a t-beam and two cut away quarter circles. The part datum can be chosen to be the lower left corner.

CH2002.m

```
partD(1,:)=[tbeam(5,8,3,1,'w','comp'),0,0,1];
partD(2,:)=[quartercircle(2,'sw','comp'),5,5,-1];
partD(3,:)=[quartercircle(2,'nw','comp'),5,0,-1];
centXD=comp(partD,'centX')
IyD=comp(partD,'Iy')
```

```
centXD =
    1.0285
IyD =
   23.6463
```

20.3   Average normal stress

What is the average normal stress at the cross section shown in Figure 20.3?

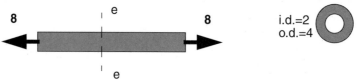

Figure 20.3   Average normal stress in a bar.

```
areaE=obeam(4,2,'area');
forceE=8;
AverageNormalE=forceE/areaE
```

```
AverageNormalE =
   0.8488
```

The next example is not as easy, see Figure 20.4. A review of the truss problems in Chapter 5 may be necessary.

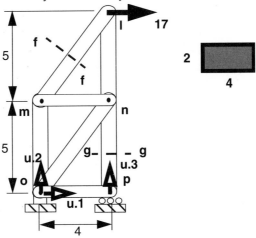

Figure 20.4   Average normal stress in truss members.

```
loadFG=[17,0,4,10];
unknownsFG=[0,0,0;DR(90),0,0;DR(90),4,0];
reactionsFG=threevector(loadFG,unknownsFG);
pointL=twovector(loadFG,[DR(-90),atan2(-5,-4)]);
magF=mag(pointL(2,:));
pointP=twovector(reactionsFG(3,:),[DR(90),DR(180)]);
magG=mag(pointP(1,:));
areaFG=rectangle(4,2,'area');
AverageNormalF=magF/areaFG
AverageNormalG=magG/areaFG
```

```
AverageNormalF =
   3.4017
AverageNormalG =
   5.3125
```

Only ten lines of code written by the user to solve the truss problem for two normal stresses. Fairly efficient by any standard.

20.4   Average shear stress

What is the average shear stress at the cross section illustrated in Figure 20.5?

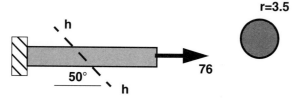

Figure 20.5   Average shear stress in a beam.

Since the cross section is on an angle, the cross section area will have to be manipulated to get the correct value. But first, the known force will have to be resolved into two new directions. Using *twovec.m* on the opposite of the original vector will give the resolution of the original (*twovec.m* gives reaction forces, so the reactions of an opposite is equal to the original.)

CH2005.m

```
forceH=[76 0 10 0];
pointH=twovector(opp(forceH),[DR(-50),DR(40)]);
shearH=pointH(1,:);
ShearMagH=mag(shearH);
areaH=circle(3.5,'area');
SlantAreaH=csc(DR(50))*areaH;
AverageShearH=ShearMagH/SlantAreaH
```

```
AverageShearH =
   0.9724
```

20.5 Using variable input

For the next problem see Figure 20.6. Start with each of the variables, *a*, *b*, *c*, set equal to 10 and solve the problem. Next change the variables one at a time so that each one of them is the only one reset to 15. Write the code so the variables need only be changed in one place to update all of the code. As with most of the work you do in MATLAB you should be creating M-files and then running them. This is particularly necessary in this situation because you will need to rewrite the third line of code for each problem, and then rerun it.

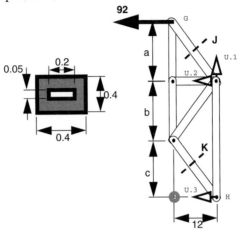

Figure 20.6   Truss of variable size.

CH2006.m

```
aVector=[10 15 10 10];
bVector=[10 10 15 10];
cVector=[10 10 10 15];
trial=1;
```

These first three lines set the variables up. For the first trial, all the values are in the first column. For the second they are in the second and so on. The variable *trial* is the only one that will have to be changed.

```
a=aVector(trial);
b=bVector(trial);
c=cVector(trial);
loadJK=[-92 0 0 (a+b+c)];
unknownDirections=[DR(90),12,b+c;DR(180),12,b+c;DR(180),12,0];
unknownForces=threevector(loadJK,unknownDirections);
PointG=twovector(loadJK, [DR(-90) atan2(-a,12)]);
magJ=mag(PointG(2,:));
PointH=twovector(unknownForces(3,:),[atan2(c,-12),DR(90)]);
magK=mag(PointH(1,:));
areaJK=rectube(0.4,0.4,0.2,0.05,'area');
AverageNormalJ=magJ/areaJK
AverageNormalK=magK/areaJK
```

```
AverageNormalJ =
  798.3811
AverageNormalK =
  399.1905
```

Now to change the values, all that need be done is to edit the variable *trial*, to equal two, or three, or four, depending on which data set needs to be run. The final answers are shown in Table 20.1

trial	a	b	c	$\sigma_{avg\text{-}j}$	$\sigma_{avg\text{-}k}$
1	10	10	10	798	399
2	15	10	10	982	599
3	10	15	10	798	319
4	10	10	15	798	393

Table 20.1   Solutions to Example 20.5

As should be expected, the stress at section *J* is dependent on only the variable *a*. This shows how many different problems can be solved quite quickly using the method of variable input.

20.6   Average normal strain

What is the strain in the cable when each of the three rocker arms is installed, the cable tightened and then the arm is rotated through 10° ccw, see Figure 20.7.

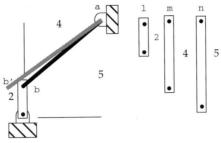

Figure 20.7   Variable input strain problem.

This problem is going to have to be set up so that each of the three different lengths of the rocker arm can be input easily. Since the deflection is going to be 10° the small angle approximation is not applicable, even if you wanted to use it.

CH2007.m

```
ArmsVector=[2 4 5];
trial=1;
arm=ArmsVector(trial);
Ax=4;
Ay=5;
Bx=0;
By=arm;
BPx=-arm*sin(DR(10));
BPy=arm*cos(DR(10));
OldLength=hyp(Ax-Bx,Ay-By);
NewLength=hyp(Ax-BPx,Ay-BPy);
NormalStrain=(NewLength-OldLength)/OldLength
```

Running the program three times, each time changing the variable *trial* to the new value, the answers are found in Table 20.1:

Length of arm	2	4	5
Normal strain	0.0599	0.1673	0.2172

Table 20.2   Solutions to Example 20.6

191

In this next example, see Figure 20.8, the cable is goes through a known amount of strain. What angle is the bar at after this strain?

p.)

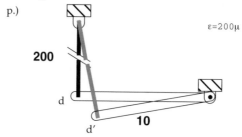

Figure 20.8   Finding deflection from strain.

CH2008.m

```
strain=200e-6;
BD=200;
CD=10;
BDP=BD*(1+strain);
CDP=CD;
BC=hyp(BD,CD);
angles=findangle(BD,CD,BC);
anglesP=findangle(BDP,CDP,BC);
theta=angles(1);
thetaP=anglesP(1);
DeltaTheta=thetaP-theta;
RD(DeltaTheta)
```

---

ans =
    0.2292

There is the code. Notice that even though this code only needs to be written once, the code was written for variable input. This is a very good habit to get into. When important numbers are placed directly into the code, that number is referred to as a "magic number." What this means, is that if you read the code later, you will not know where that number came from, it is just "magic." A better method is the way it was done here. If the fourth line were to read
```
>>BDP=BD*(1+200e-6);
```
It is not as readily apparent what is going on. After the first three lines, the only two numerals to appear in the code are referring to a position within a vector of numbers.

20.7   Average shear strain

What is the average shear stress at the upper left corner shown in Figure 20.9?

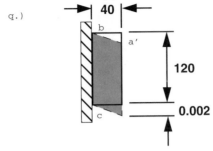

Figure 20.9   Shear strain.

CH2009.m

```
AngleAPBC=atan2(40,0.002);
ShearStrain=DR(90)-AngleAPBC
```

```
ShearStrain =
5.0000e -05
```

20.8   Stress manipulation

Making the assumption of plane stress, what are the principle stresses shown in Figure 20.10?

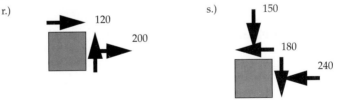

Figure 20.10   Principle stress example

```
StressStateR=[200,0,120];
StressStateS=[-240,-150,-180];
PrincipleStressR=pristress(StressStateR,'plane stress')
PrincipleStressS=pristress(StressStateS,'plane stress')
```

```
PrincipleStressR =
 -56.2050        0 256.2050
PrincipleStressS =
-380.5398  -9.4602         0
```
Remember the notation convention:

Stress state vector:

$$\left[\text{normal stress x, normal stress y, shear stress x-y}\right]$$

For the same two examples, find the principle planes. Continuing with the variables from above:

>>PrinciplePlanesR=ppstress(StressStateR)
```
PrinciplePlanesR =
    0.4380
    2.0088
```
>>PrinciplePlanesS=ppstress(StressStateS)
```
PrinciplePlanesS =
    0.6629
    2.2337
```
Remember, these answers are in radians. To convert them to degrees, use the *RD* function.
>>RD(PrinciplePlanesR)
```
  25.0972
 115.0972
```
>>RD(PrinciplePlanesS)
```
  37.9819
 127.9819
```

20.9   Strain manipulation

Making the assumptions of plane strain, what are the principle strains in these situations illustrated in Figure 20.11?

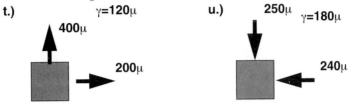

Figure 20.11   Principle strains example.

CH2011.m

```
StrainStateT=[200 400 120]*1e-6;
StrainStateU=[-240 -250 180]*1e-6;
PrincipleStrainT=pristrain(StrainStateT,'plane strain')
PrincipleStrainU=pristrain(StrainStateU,'plane strain')
```

```
PrincipleStrainT =
   1.0e-03
        0    0.1834    0.4166
PrincipleStrainU =
   1.0e-03 *
  -0.3351   -0.1549          0
```
Strain state vector:

$$\left[\text{normal strain x   normal strain y   shear strain x-y}\right]$$

For the same two examples, find the principle planes. Continuing with the variables from above:

```
>>PrinciplePlanesT=ppstrain(StrainStateT)
PrinciplePlanesT =
  -0.2702
   1.3006
>>PrinciplePlanesU=ppstrain(StrainStateU)
PrinciplePlanesU =
   0.7576
   2.3284
```

Remember, these answers are in radians. To convert them to degrees, use the *RD* function.

```
>>RD(PrinciplePlanesT)
  -15.4819
   74.5181
>>RD(PrinciplePlanesU)
   43.4101
   133.4101
```

20.10 Stress transformation

What is the stress state on the angled planes shown in Figure 20.12?

v.)

w.)

Figure 20.12   Stress transformation.

CH2012.m

```
StressStateV=[800 0 700];
StressStateW=[-240 180 -400];
[NormalV,ShearV]=stresstr(StressStateV,DR(180-60))
[NormalW,ShearW]=stresstr(StressStateW,DR(45))
```

```
NormalV =
    1.2062e+03
ShearV =
    3.5898
NormalW =
  370.0000
ShearW =
 -210.0000
```

20.11   Strain transformation

What is the strain state on the reoriented element as illustrated in Figure 20.13?

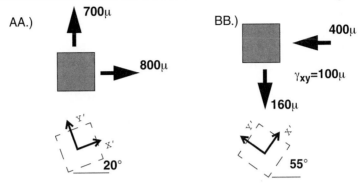

Figure 20.13   Stress transformation.

CH2013.m

```
StrainStateAA=[800 700 0];
StrainStateBB=[-400 160 100];
NewStrainStateAA=straintr(StrainStateAA,DR(20))
NewStrainStateBB=straintr(StrainStateBB,DR(55))
```

```
NewStrainState =
  788.3022711.6978 -64.2788
NewStrainStateBB =
  22.7503-262.7503 492.0259
```

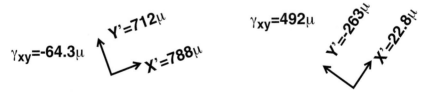

20.12   Strain Rosettes

A strain rosette is set up on piece of equipment made out of tool steel. When the test is executed, the following data is collected:

$$\varepsilon_0 = 97\mu \qquad \varepsilon_{45} = 458\mu \qquad \varepsilon_{90} = 424$$

What are the principle stresses in this material in SI units?

```
>>EpsilonCC=[97 458 424]*1e-6;
>>ThetaCC=DR([0 45 90]);
>>StrainStateCC=rosette(EpsilonCC,ThetaCC)
StrainStateCC =
1.0e-03 *
   0.0970    0.4240   -0.3950
>>ECC=matprop('tool steel','E','SI');
>>PoissonsCC=matprop('tool steel','Poissons','SI');
>>StressStateCC=strain2stress(StrainStateCC,ECC,PoissonsCC)
StressStateCC =
   0.0518  0.1014    0.0299
>>PrincipleStressesCC=pristress(StressStateCC,'plane
stress')
PrincipleStressesCC =
      0   0.0378   0.1155
```

In this book, units have been intentionally used sparingly. For this example what units are the answers in? The units of the Young's modulus are in GPa, therefore so are the units of the stresses.

$$0.1155 \text{ GPa} = 115.5 \text{ MPa} \ .$$

20.13   Mohr's circle: stress

Draw the Mohr's circle representation of the following stress states shown in Figure 20.14. Assume plane stress. For the second problem, represent the diagonal plane on the drawing also. See Figures 20.15 and 20.16 for Mohr's representations.

Figure 20.14   Stress states for Mohr's.

CH2014.m

```
StressStateDD=[17 -20 18];
StressStateEE=[10 100 58];
figure(1)
mohrs(StressStateDD,'plane stress')
figure (2)
mohrs(StressStateEE,'plane stress',DR(45))
```

The figure commands simply ensure that a new window is opened for the pictures. This becomes necessary when multiple windows are to be opened. Consult the MATLAB on-line help.

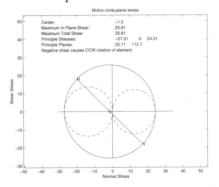

Figure 20.15   Mohr's circle representation.

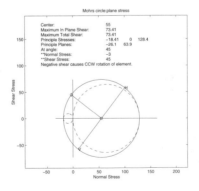

Figure 20.16   Mohr's circle representation.

20.14 Mohr's circle: strain

Draw the Mohr's circle representation of the strain states shown in Figure 20.17. See Figures 20.18 and 20.19 for representations.

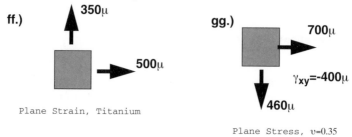

ff.) 350μ

500μ

Plane Strain, Titanium

gg.) 700μ

γ_{xy}=-400μ

460μ

Plane Stress, υ=0.35

Figure 20.17   Strain states for Mohr's circle.

CH2015.m

```
StrainStateFF=[500,350,0];
StrainStateGG=[700,460,-400];
PoisonsFF=matprop('titanium','Poissons','SI');
figure(1)
mohrs2(StrainStateFF,'plane strain',PoisonsFF)
figure(2)
mohrs2(StrainStateGG,'plane stress',0.35)
```

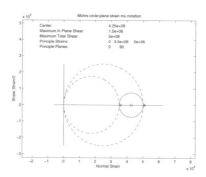

Figure 20.18   Mohr's for strain.

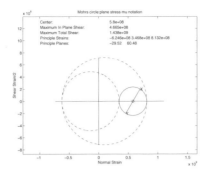

Figure 20.19   Mohr's strain.

20.15 Stress strain conversions

Earlier in these examples the stress strain conversion was shown by way of the strain rosette, Example 20.12. But this can also be done in reverse. What is the strain state that would occur due to the stress state that is applied as shown in Figure 20.20?

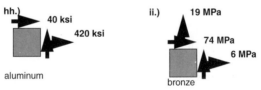

Figure 20.20   Stress strain conversions.

CH2016.m

```
StressStateHH=[420 0 40]*1e3;
StressStateII=[6 19 74]*1e3;
EHH=matprop('aluminum','E','US');
PoissonsHH=matprop('aluminum','Poissons','US');
EII=matprop('bronze','E','SI');
PoissonsII=matprop('bronze','Poissons','SI');
StrainStateHH=stress2strain(StressStateHH,EHH,PoissonsHH)
StrainStateII=stress2strain(StressStateII,EII,PoissonsII)
```

```
StrainStateHH =
1.0e+04 *
   3.9623   -1.3868    1.0189
StrainStateII =
1.0e+03 *
  -0.0045    0.1647    1.9254
```

Looking at the units on these problems, the Young's modulus value must be in the same units as the given stresses. This unit change was done in the first two lines of code. When these units are done properly, the strain will come out unitless.

20.16   A complete example

This example will start with data from a strain rosette, and then find all of the following data with it:
- Strain state
- Principle strains
- Principle planes of strains
- Strain state on an arbitrary plane
- Draw the Mohr's circle for strain
- Stress state
- Principle stresses
- Princip.e planes of stress
- Stress state on an arbitrary plane
- Draw the Mohr's circle for stress
  The strain rosette readings are:

$$\varepsilon_0 = 97\mu \qquad \varepsilon_{45} = 458\mu \qquad \varepsilon_{90} = 424$$

The material being studied is structural steel. Use SI units under the assumption of plane stress. The arbitrary plane should be 30° ccw from right horizontal.

```
Epsilons=[100 150 45]*1e-6;
Thetas=DR([0 45 90]);
Youngs=matprop('structural steel','E','SI');
Poissons=matprop('structural steel','Poissons','SI');
Angle=DR(30);
StrainState=rosette(Epsilons,Thetas)
PrincipleStrains=pristrain(StrainState,'plane stress')
PrinciplePlanesStrain=ppstrain(StrainState)
NewStrainState=straintr(StrainState, Angle)
figure(1)
mohrs2 (StrainState,'plane stress',Poissons,Angle)
StressState=strain2stress(StrainState,Youngs,Poissons)
PrincipleStresses=pristrain(StressState,'plane stress')
PrinciplePlanesStress=ppstress(StressState)
[NormalStress ShearStress]=stresstr(StressState,Angle)
figure(2)
mohrs(StressState,'plane stress',Angle)
```

This code was designed using the variable input method. If ever it were desirable to run the code for a new set of rosette data on a different type of material with a different arbitrary angle of interest, all of that could be changed just once in the top lines and the rest of the code would still work.

To add notes to the Mohr's circle diagrams see Figures 20.21 and 20.22, use the command *gtext*. For example, to note that the units are GPa you would first make sure the correct figure was active, and then run the command. When the *gtext* command is running, cross hairs will appear so that the lower left corner of the text can be placed on the drawing.

```
>>figure(2)
>>gtext ('All stress units are GPa')
```

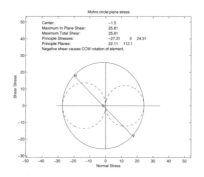

Figure 20.21   Mohr's circle for strain.

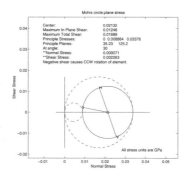

Figure 20.22   Mohr's circle for stress.

21

Axial Loading

21.1 Introduction

There are many different ways that a stress can be induced upon a mechanical element. Thus far we have learned how to manipulate stress and strain states with little regard of how to find these states from the applied loads.

This chapter will discuss stresses and strains that are induced due to centric, or axial loading. A total of six loading types will be covered: Axial, thermal, torsional, bending, shear and pressure vessels. These loading types can then be combined using the principle of superposition to find the stress state of a differential element on a loaded body. This stress state can then be manipulated as covered in the previous chapters.

This chapter is also the first to find forces in situations that are statically indeterminate. That is to say that the equilibrium equations are not sufficient to determine the reactions.

21.2 Problem

If a bar is supported at two ends and is subjected to an axial load see Figure 21.1, then the problem is no longer statically determinant. Material properties will have to be invoked to find the support reactions.

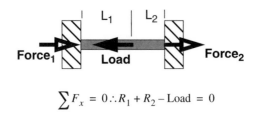

$$\sum F_x = 0 \therefore R_1 + R_2 - \text{Load} = 0$$

Figure 21.1   Statically indeterminate axial loading.

21.3   Theory

The axial loading covered here is an elaboration of the material discussed in Chapter 10 "Average Normal and Shear Stresses." The stress induced in an axially loaded member is:

$$\sigma = \frac{\text{Force}}{\text{Area}} \qquad \text{Eq. 21.1}$$

The mathematical definition of strain is:

$$\varepsilon = \frac{\Delta \text{Length}}{\text{Length}} \qquad \text{Eq. 21.2}$$

The material constant that relates stress and strain is Young's modulus, E, it can be found using the routine *matprop.m*.

$$E = \frac{\sigma}{\varepsilon} \qquad \text{Eq. 21.3}$$

These can be combined and rearranged into the forms:

$$\Delta \text{Length} = \frac{\text{Force} \times \text{Length}}{\text{Area} \times E} = \text{Force} \times K \qquad \text{Eq. 21.4}$$

$$\text{Force} = \frac{\Delta \text{Length} \times E \times \text{Area}}{\text{Length}} = \frac{\Delta \text{Length}}{K} \qquad \text{Eq. 21.5}$$

Looking back at Figure 21.1, it can be seen that the equilibrium equation yields one equation and two unknowns, R_1 and R_2. This is not a solvable system. However, the displacements caused by the forces must also sum to zero. This will give a second equation without introducing any new unknown quantities. The system will then be solvable.

$$\Delta \text{Length}_1 + \Delta \text{Length}_2 = 0 \qquad \text{Eq. 21.6}$$

Rearranging that gives the equation:

$$\frac{\text{Force}_1 \times \text{Length}_1}{\text{Area}_1 \times E_1} + \frac{\text{Force}_2 \times \text{Length}_2}{\text{Area}_2 \times E_2} = 0 \qquad \text{Eq. 21.7}$$

In the above equation it should be noticed that the force the bar sees differs depending on if the point of interest lies to the left, region one, or the right, region two, of the applied load. The force seen in the region one is equal to the value of reaction one. The force in region two is equal to the value of reaction two. This can be combined with the equilibrium equation:

$$\sum F_x = 0 \therefore \text{Force}_1 + \text{Force}_2 - \text{Load} = 0 \qquad \text{Eq. 21.8}$$

Equation 21.7 and Equation 21.8 give a solvable system. Now for the task of finding the easiest method for solving the system of governing equations. The force method of analysis is the one to be used in this book.

The force method of analysis, or the flexibility method, temporarily removes one of the supporting forces to allow the bar to deform in a statically determinant way. Then the supporting force can be calculated by asking what it must be to cause the bar to be deformed to the final shape. Finally, the other support reaction force can be calculated through simple equilibrium.

There are two general cases that can complicate the situation, see Figure 21.2, both of which will be covered in the next section. The two situations are either:

- Two or more members of differing stiffness, $\dfrac{\text{Length}}{\text{Area} \times \text{E}}$, all deflect differently under the same force.
- Two or more members of differing stiffness all carry a different load but have the same deflection.

In both of these situations, using a ratio of the stiffness of the different members, the required information can be found.

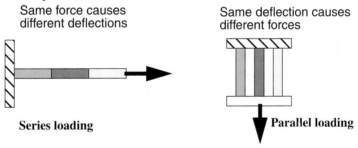

Same force causes
different deflections

Same deflection causes
different forces

Series loading

Parallel loading

Figure 21.2   Two cases of indeterminate loading.

21.4   Output

Statically indeterminate, no complications: What are the reactions on the bar   illustrated in Figure 21.3?

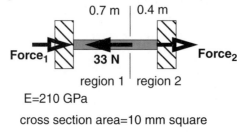

Figure 21.3   Axially loaded member.

For this problem *Force₂* will be temporarily removed to see how the bar deforms. The deformation will be only in region one because there is no stress applied to region two. Using Equation 21.4 with all units in meters and Newtons:

CH2101.m

```
af=-33;%Newtons
LengthRegionOne=0.7;%meters
LengthTotal=1.1;%meters
Area=(10e-3)^2;%meters^2
E=210e9;%Pascals
Deformation=(af*LengthRegionOne)/(Area*E)%meters
```

```
Deformation =
-1.1000e-06
```

For that deformation to be counteracted, *Force₂* must cause a deformation equal but **opposite** to it. This time, the loading force can be ignored. Without the force in the middle to break the bar into two regions, the length to be used in the calculations is the length of the entire bar. Using Equation 21.5 with the same units:

```
>>ForceTwo=-Deformation*E*Area/LengthTotal
ForceTwo =
   21.0000
>>ForceOne=-(af+ForceTwo)
ForceOne =
   12.0000
```

Now that the forces are known, it is very simple to go to the stresses by dividing by the areas the forces work on:

```
>>StressOne=ForceOne/Area
StressOne =
1.2000e+05
>>StressTwo=ForceTwo/Area
StressTwo =
2.1000e+05
```

This example is a good one, but it is not very general. It also has made some unseen assumptions, the biggest of these is that the Young's modulus and the area are unchanged along the length of the bar. This will cause some complications in the general case.

Series loading: What if there were more loads in the middle of the bar, or the area or material changed from region to region? What if there is a gap into which the bar can expand before it touches the second support wall? For these reasons a more general template is developed here.

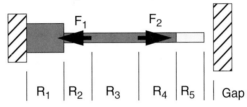

Figure 21.4   General case of axial loading with constant force.

For this general case, see Figure 21.4 a new region is formed whenever there is

- Force introduced
- A change in cross sectional area
- A change in material properties

Basically, a region is defined as an area with constant cross section, constant material properties, and is subjected to the same force throughout. Because of these constant properties, the stiffness in a region is constant throughout the region.

The values for the properties of force, length, area, and Young's modulus must be put into matrixes to keep them organized. Putting numbers with the general case above will clarify this process, see Figure 21.5.

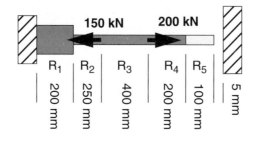

Figure 21.5   Numerical values for the general case.

Ignore for now the reaction force that will occur on the right end after the gap is closed. The reaction force at the left wall must be -50 kilo-Newtons to equal out the applied loads. Working from the left, the force in region one must be 50 kilo-Newtons tension. Tension is a positive value, compression is negative. Region two sees no force change from region one, only an area change. Region three increases the tension to 200 kN. This force is dropped down to zero for the remaining two regions. Putting these and the other values for length, area, and Young's modulus into vector form:

```
>>Force=[50 50 200 0 0]*1e3;%Newtons
>>Length=[200 250 400 200 100]*1e-3;%meters
>>Radii=[6 2.5 2.5 2.5 2.5]*1e-3;%meters
>>Youngs=[210 210 210 210 180]*1e9;%Pascals
>>Gap=5e-3;%meters
>>AppliedLoads=[-150 200]*1e3;%Newtons
>>Area=(Radii.^2)*pi;%meters^2
```

Now these data vectors can be manipulated just as the scalars were in the previous example. However, in this case, vectors are being multiplied and divided. To tell MATLAB that operations are not matrix multiplications, but rather a series of scalar multiplications, the array multiplication and division must be used. This is done by placing a period before the normal operation. Do a help on "ops" within MATLAB for more information.

```
>>Deflection=(Force.*Length)./(Area.*Youngs)
Deflection =
    0.0004    0.0030    0.0194         0         0
```

This vector is interpreted to mean that region one increased in length by 0.8 mm, region two by 2.4 mm, region three by 15.2 mm and the remaining regions were not deformed. Next the total deflection can be found:

```
>>TotalDeflection=sum(Deflection)
TotalDeflection =
    0.0229
```

Next the original loadings on the bar are ignored and the force supplied by the left-hand wall must be calculated. The force must be sufficient to deform the bar back through the distance it expanded. Actually, the wall does not take effect giving a reaction force until the bar has expanded through the distance of the gap.

```
>>ForcedDeflection=TotalDeflection-Gap
ForcedDeflection =
    0.0179
```

Knowing the length that must be deflected by the reaction force is not enough. The contraction of the entire bar is known, it is also known to be the sum of the contractions of each section. Each of the regions will be contracted in a different ratio of the entire deflection depending on its stiffness, $K = \dfrac{\text{Length}}{\text{Area} \times \text{E}}$. A stiffer region is going to take a smaller percentage of the contraction than the less stiff regions are. The percentage of the deflection a region will take up is:

$$\text{Percentage}_n = \frac{K_n}{\sum K_N} \qquad \text{Eq. 21.9}$$

Knowing that a certain region will take up a fixed percentage of the deflection allows the force in that region to be found with Equation 21.5. The force applied is the same in all regions, and in fact is equal to the force supplied by the right-hand reaction.

```
>>Stiffness=Length./(Area.*Youngs);
>>Percentage=Stiffness./sum(Stiffness);
>>RegionalDeflection=ForcedDeflection.*Percentage;
>>ReactionRight=-RegionalDeflection.*Youngs.*Area./Length

ReactionRight =
   1.0e+04 *
  -7.3517   -7.3517   -7.3517   -7.3517   -7.3517
```

This *ReactionRight* is actually a vector with the reaction repeated through the vector once for each region. One time will be sufficient, so it will be saved in a new scalar and the problem will be completed.

```
>>GapReaction=ReactionRight(1);
>>WallReaction=-sum(AppliedLoads)-GapReaction
WallReaction =
2.3517e+04
```

Here is the entire M-File gathered together without the commentary:

CH2102.m

```
Force=[50 50 200 0 0]*1e3;%Newtons
Length=[200 250 400 200 100]*1e-3;%meters
Radii=[6 2.5 2.5 2.5 2.5]*1e-3;%meters
Youngs=[210 210 210 210 180]*1e9;%Pascals
Gap=5e-3;%meters
AppliedLoads=[-150 200]*1e3;%Newtons
Area=(Radii.^2)*pi;%meters^2
Deflection=(Force.*Length)./(Area.*Youngs);
TotalDeflection=sum(Deflection);
ForcedDeflection=TotalDeflection-Gap;
Stiffness=Length./(Area.*Youngs);
Percentage=Stiffness./sum(Stiffness);
RegionalDeflection=ForcedDeflection.*Percentage;
ReactionRight=-RegionalDeflection.*Youngs.*Area./Length;
GapReaction=ReactionRight(1)
WallReaction=-sum(AppliedLoads)-GapReaction
```

This does not look too complicated, in fact 6 of the 15 lines are used to input the data. It is written in the variable input format, so different situations can be ran through it.

It is entirely possible when running different sets of data through that the *GapReaction* will be in the positive direction. This would mathematically mean that the right hand wall was applying a tension force to the bar. In physical reality, this is not possible. All this means is that the bar never expanded far enough to contact the wall. This would then imply that the problem is statically determinant and the *WallReaction* is simply equal to the opposite of the sum of the applied forces.

Finding the stresses in each area would be a bit more complicated because not only does the force vary from region to region, but so does the area. The actual force in each region has not been gathered yet. This type of work can also be applied to torsional loading, and will be covered on its own in Chapter 23. The regional forces that were gathered in the beginning of the routine are not valid for a stress analysis because those forces temporarily ignored one of the support reactions.

This is the most common form in which to see this type of problem. Often times the normal stresses are calculated once the reaction forces are known, but that has already been covered. Rarely are the reaction forces known and the loading to be discovered.

Parallel loading, finding displacement: Another situation that can occur to complicate the problem is when all the members deflect through the same distance because of an applied load. However, due to the differing stiffness in each of the members, more or less of the force is carried by each member.

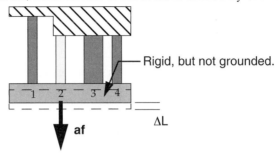

Figure 21.6   General case of axial loading with constant deflection.

In Figure 21.6, the lower horizontal bar is rigid, and remains horizontal. Each of the supporting bars has a length, cross sectional area, and material that is independent of the others. What is the *af* required for the horizontal bar to be displace ΔL.

First as with the other problem, the characteristics of each member should be gathered into vectors:

```
>>Lengths=[0.7 0.5 0.5 0.5]; %meters
>>Areas=[0.03 0.03 0.06 0.03].^2*pi; %getting area from
radii
>>Youngs=[210 180 210 210]*1e9; %Pascals
>>DeltaL=0.008;%meters
```

Next the force required to deflect each of the bars through the same distance must be calculated using Equation 21.5:

```
>>Forces=DeltaL*Youngs.*Areas./Lengths
Forces =
1.0e+07 *
    0.6786   0.8143   3.8001   0.9500
```

This vector can be read as: "The force in the first member is 6.786 MN, the force in the second member is 8.143 MN, the force is the third member is 38.001 MN, and the force in the final member is 9.500 MN."

These answers seem reasonable. If the fourth bar is considered the datum to which the others are compared then the force to deflect the first bar should be less because the bar is longer, just as the force to deflect the second bar should be less because the material is less stiff. The third bar should take more force because it is of a larger diameter. The results reflect this intuitive knowledge, they pass this logical test.

Finally, the force required to move the horizontal bar is going to simply be the sum of the individual forces.

```
>>Load=sum(Forces)
Load =
6.2430e+07
```

The stresses are easy to find in this situation also:

```
>>Stresses=Forces./Areas
1.0e+09 *
    2.4000    2.8800    3.3600    3.3600
```

The M-file gathered together without the commentary:

CH2103.m

```
Length=[0.7 0.5 0.5 0.5]
Area=[0.03 0.03 0.06 0.03].^2*pi; %getting area from radii
Youngs=[210 180 210 210]*1e9;
DeltaL=0.008;
%%%% End Set up Variables %%%%
Forces=DeltaL*Youngs.*Area./Lengths;
Load=sum(Forces)
Stresses=Forces./Area
```

Again, it is true that setting up the variables is a major portion of the problem.

Parallel loading, finding forces: One more situation that can occur to this type of problem has more variations than the others. What if the same setup was used, but the force was known and the deflection was unknown? This would require the stiffnesses to be used to figure out the ratio in which the forces are distributed across the many bars. Use the answer from the previous problem and see if the same deflection is calculated.

If you are following these examples at a terminal, be sure to *clear* the variable between different programs so the answers from the last program do not hold values when you think they are not yet defined. Clearing and then starting with the same variable setup:

```
>>clear %undefining all of the variables
>>clc %clearing the command window of text
>>Length=[0.7 0.5 0.5 0.5];
>>Area=[0.03 0.03 0.06 0.03].^2*pi; %getting area from radii
>>Youngs=[210 180 210 210]*1e9;
>>Load=6.2430e+07;
```

Next the stiffnesses need to be calculated and the percentage given to each bar defined:

```
>>Stiffness=Length./(Area.*Youngs);
>>Percentage=(1./Stiffness)./sum(1./Stiffness);
>>Forces=Percentage*Load
Forces =
1.0e+07 *
    0.6786   0.8143   3.8001   0.9500
```

So far so good, the values for the individual forces carried by the members are the same as the previous example. Now to find the deflection based on these forces.

```
>>Deflectons=Forces.*Length./(Area.*Youngs)
Deflectons =
    0.0080   0.0080   0.0080   0.0080
```

As should be expected, the deflections in the bars are all equal to each other and are equal to ΔL. The last step is to capture ΔL in it's own variable.

```
>>DeltaL=Deflectons(1)
DeltaL =
    0.0080
```

These examples work in either direction, from deflection to load or from load to deflection.

CH2104.m

```
clear %undefining all of the variables
clc %clearing the command window of text
Length=[0.7 0.5 0.5 0.5];
Area=[0.03 0.03 0.06 0.03].^2*pi; %getting area from radii
Youngs=[210 180 210 210]*1e9;
Load=6.2430e+07;
Stiffness=Length./(Area.*Youngs);
Percentage=(1./Stiffness)./sum(1./Stiffness);
Forces=Percentage*Load
Deflectons=Forces.*Length./(Area.*Youngs)
DeltaL=Deflectons(1)
```

21.5   Summary

Statically indeterminate axially loaded problems fall into two major categories:
1.   The deflection of the structure is spread among many different members in a set ratio based on their stiffnesses, see Figure 21.7

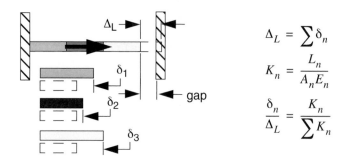

$$\Delta_L = \sum \delta_n$$

$$K_n = \frac{L_n}{A_n E_n}$$

$$\frac{\delta_n}{\Delta_L} = \frac{K_n}{\sum K_n}$$

Figure 21.7   Summary: equal forces.

2. The force on a structure is carried among the many different members in a set ratio based on their stiffnesses see Figure 21.8.

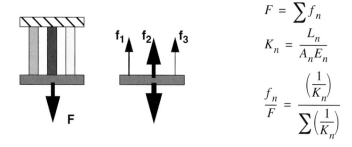

$$F = \sum f_n$$

$$K_n = \frac{L_n}{A_n E_n}$$

$$\frac{f_n}{F} = \frac{\left(\frac{1}{K_n}\right)}{\sum \left(\frac{1}{K_n}\right)}$$

Figure 21.8   Summary: equal displacements.

Knowing the distribution of forces or the distribution of deflections allows for the single regions to be solved with the simple equations:

$$\Delta \text{Length} = \frac{\text{Force} \times \text{Length}}{\text{Area} \times E} = \text{Force} \times K$$

$$\text{Force} = \frac{\Delta \text{Length} \times E \times \text{Area}}{\text{Length}} = \frac{\Delta \text{Length}}{K}$$

22

Thermal Loading

22.1 Introduction

The next type of loading that will be discussed is thermal loading. The types of problems that are generally encountered with thermal loading are fairly similar to the ones discussed in the previous chapter. Thermal stresses and strains are induced when a body is exposed to a change in temperature. It will react to this thermal change in a way that is predictable based on the materials coefficient of thermal expansion.

22.2 Problem

What are the stresses and strains induced in a body due to changes in temperature? See size change illustrated in Figure 22.1.

Figure 22.1   Size change due to temperature increase.

22.3   Theory

The strain induced in a material due to a change in temperature is a material property just like Poisson's ratio. The value that defines this property is called the coefficient of thermal expansion. It can be found using the routine *matprop.m* or in a standard reference book. The governing equation is very simple:

$$\varepsilon_{thermal} = \alpha(\Delta T) \qquad\qquad \text{Eq. 22.1}$$

22.4   Output

There are a few variations on the problem of thermal loading, first the simplest case will be solve. Then a more general case that includes the first example as a subset will be solved. Finally a more unique variation will be solved.

Simplest thermal loading: What is the force induced in the bar, shown in Figure 22.2, due to the thermal expansion?

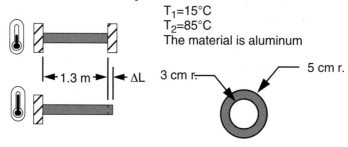

$T_1 = 15°C$
$T_2 = 85°C$
The material is aluminum

1.3 m ΔL 3 cm r.   5 cm r.

Figure 22.2   Simple thermal expansion problem.

The method of solving this is nearly identical to the method of solving the axially load problems of the previous chapter. First one of the supports is ignored to see what the deflection of the bar would be. Then the amount of force that must be supplied by the wall to move the bar back to it's original length is found. Using Equation 22.1 to find ΔL:

CH2201.m

```
DeltaT=85-15;% C
Length=1.3;% meters
Alpha=matprop('aluminum','thermal expansion','SI');% 1/C
E=matprop('aluminum','E','SI')*1e9;% Pascals
Strain=Alpha*DeltaT;% unitless
DeltaL=Length*(1+Strain);% meters
Area=obeam(0.05, 0.03, 'area');% meters^2
```

A force is required to force the bar back through the distance ΔL. Remembering Equation 21.5:

$$\text{Force} = \frac{\Delta\text{Length} \times E \times \text{Area}}{\text{Length}}$$

```
>>Force=DeltaL*E*Area/Length %Newtons
Force =
9.2008e+07
```

The force can be read as 92 MN.

To get the stress imposed by the temperature change, divide by the area:

```
>>Stress=Force/Area
Stress =
7.3218e+10
```

This first case was simple enough to do. All other thermal stress problems are variations on this theme.

General case of expansion in series: A slightly more complicated situation would have different regions in the bar, each with a different material or area, see Figure 22.3. The next M-file will solve the case where there are *n* regions to the bar. You may recall this technique from the previous chapter:

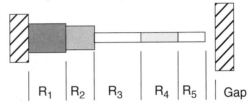

Figure 22.3   Thermal loading on a multiregion bar.

As always, the first step is to gather the data into vectors for easy manipulation. Though the values are not shown on the figure:

CH2202.m

```
Lengths=[250 200 300 250 150]*1e-3; %meters
Radii=[30 25 15 15 15]*1e-3; %meters
Youngs=[210 200 190 195 190]*1e9; %Pascals
Alphas=[12 13 15 14 15]*1e-6; % 1/C
Gap=5e-3; %meters
DeltaT=80; %C
```

The next step is to calculate the areas based on the radii, and the strain based on the temperature change. This strain will be used to discover the deflection that would occur if the left-hand wall is neglected.

219

```
Areas=pi*(Radii.^2); %meters^2
Strains=Alphas*DeltaT; %unitless
Deflections=Strains.*Lengths; %meters
TotalDeflection=sum(Deflections); %meters
```

Next, the amount that the beam wishes to expand beyond the wall is calculated. Then the stiffness ratios are calculated to find the amount of the deflection that is taken by each of the different regions. Knowing that the same force is compressing all of the sections, and knowing what the deflection of each section is, allows the force to be found. Using the equilibrium equations, the remaining reaction at the left hand wall can be found.

```
ForcedDeflection=TotalDeflection-Gap; %meters
Stiffness=Lengths./(Areas.*Youngs); %meters/Newton
Percentages=Stiffness./sum(Stiffness); %unitless
RegionalDeflection=ForcedDeflection.*Percentages; %meters
ReactionRight=RegionalDeflection.*Youngs.*Areas./Lengths;%Newton
GapReaction=ReactionRight(1) %Newtons
GapReaction =
-6.1234e+05
WallReaction=-GapReaction %Newtons
WallReaction =
6.1234e+05
```

Since the force is the same throughout the body, to find the stress all that remains to be done is to divide by the areas of the different regions

```
Stresses=WallReaction./Areas %Pascals
Stresses =
    1.0e+08 *
    2.1657  3.1186    8.6628    8.6628    8.6628
```

These answers can be read as: "The first region is experiencing 217 MPa of stress, the second is experiencing 312 MPa and the last three are experiencing 866 MPa."

Notice with this problem, that the order of the regions is of no importance, also notice that the gap could be anywhere. It could be at either end or it could be between two regions. Knowing that, several other variations of this problem to be solved. In fact, if there were two gaps somehow involved, the two could simply be summed and be used as one.

Notice also with this code, that if there was only one region, that is the bar was homogenous throughout, the code would still work. So, this code could replace the M-file presented for the first example. The procedure in this problem is more complicated than is needed to solve the first problem, but that is often the case with code written to solve general cases.

Parallel Expansion

A more difficult problem might be to find the final length of the sleeve and the force in the sleeve in this illustrated assembly if the assembly starts out just snug and then is heated, see Figure 22.4.

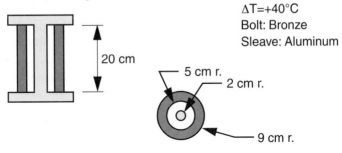

Figure 22.4   Bolt and sleeve configuration in thermal stress.

The first step is to gather the data into the proper variables:

CH2203.m

```
DeltaT=40;
LengthStart=0.20; %meters
AreaSleeve=obeam(0.09, 0.05, 'area'); %meters^2
AreaBolt=circle(0.02, 'area');%meters^2
AlphaBolt=matprop('bronze', 'thermal expansion', 'SI');%1/C
AlphaSleeve=matprop('aluminum', 'thermal expansion', 'SI');%1/C
EBolt=matprop('bronze', 'E', 'SI')*1e9;%Pascals
ESleeve=matprop('aluminum', 'E', 'SI');%Pascals
```

The next is to find the length the bolt and the sleeve would expand to if not constrained by each other:

```
StrainFreeBolt=DeltaT*AlphaBolt;%unitless
StrainFreeSleeve=DeltaT*AlphaSleeve;%unitless
LengthFreeBolt=(1+StrainFreeBolt)*LengthStart;%meters
LengthFreeSleeve=(1+StrainFreeSleeve)*LengthStart;%meters
DeltaLengths=LengthFreeSleeve-LengthFreeBolt;%meters
```

This length difference is the distance by which the sleeve wishes to expand beyond the bolt. There will be a compromise between the two, the bolt will expand farther than it wants to, and the sleeve will expand less than it wants to. The ratio of the amount of length each of the members is forced through is defined by their stiffnesses:

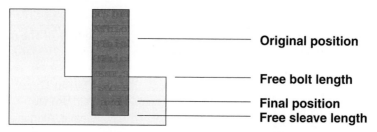

Original position

Free bolt length

Final position
Free sleave length

```
KBolt=LengthFreeBolt/(AreaBolt*EBolt);%meters/Newtons
KSleeve=LengthFreeSleeve/(AreaSleeve*ESleeve);%meters/Newton
DeltaBolt=DeltaLengths*KBolt/(KBolt+KSleeve);%meters
DeltaSleeve=-DeltaLengths*KSleeve/(KBolt+KSleeve);%meters
FinalLength=LengthFreeBolt+DeltaBolt %meters
FinalLength =
0.20013600000012
```

The long form of the answer is shown here for comparison to the answer arrived at in a similar problem in Section 29.3

Knowing the distance that each member is forced through by the other allows the forces to be calculated. Since the two members only interact with each other, the forces carried by them should sum to zero.

```
ForceBolt=DeltaBolt*EBolt*AreaBolt/LengthFreeBolt %Newtons
ForceBolt =
    0.0295
```

```
ForceSleeve=DeltaSleeve*ESleeve*AreaSleeve/LengthFreeSleeve %N
ForceSleave =
   -0.0295
```

Positive value for the bolt means it is in tension, whereas the negative value on the sleeve means it is in compression. It is possible that if a new set of data were run through the routine by editing the first few lines, that the signs would be reversed. This would make no sense in the problem because the two members could not interact with each other in that way. This type of answer would imply that the bolt expanded farther than the sleeve, and so there would be no forces induced.

To find the stresses, simply divide by the areas:

```
StressBolt=ForceBolt/AreaBolt % Pascals
StressBolt =
    0.0613
```

```
StressSleeve=ForceSleeve/AreaSleeve %Pascals
StressSleave =
   -0.0175
```

Notice in this example, the values for the lengths were not gathered into a single data vector, but the lengths were kept in two different scalars. Because of this each calculation had to be done twice. With the method of storing all of the values in one vector, the calculations could be done all at once with vector mathematics. It is usually better to do vector mathematics when possible. It makes the code more compact, and in large problems with hundreds or thousands of data values the calculations are appreciably quicker.

A more general variation on this problem might be where there is not just one sleeve, but several. This variation has a very elegant solution that is applicable for n sleeves, and will be covered in Chapter 32 "Simultaneous Equations".

22.5   Summary

There are two major variations in this type of thermal loading. Either the different materials expand in series, where all of the deflections add together, as in the case of a beam made of several different materials and cross-sections. The simplest case of this involves a bar of a single material.

The other variation on this theme is when the different materials expand in parallel. The example of this is a bolt and sleeve configuration. These types of problems bear much resemblance to the problems of axial loading. Consequently, they are solved in much the same way.

23

Torsional Loading

23.1 Introduction

The next type of loading that will be discussed is torsional loading. Through the use of quasi-static analysis the work developed here can be just as applicable to drive shafts as they are to cantilevered beams.

23.2 Problem

When an element is subjected to torsional loading, find the deflections and strains that it undergoes.

23.3 Theory

The equations that govern this type of loading are:

$$\tau = \frac{Tr}{J}$$ Eq. 23.1

$$\theta = \frac{TL}{JG}$$ Eq. 23.2

These represent the shear stress and the deflection where:
- T = Torque
- r = Radius to the point of interest
- J = The polar moment of inertia
- L = Length of the torqued section
- G = Modulus of rigidity

Rearranging the equations to find torque as a function of the other variables:

$$T = \frac{\tau J}{r}$$
Eq. 23.3

$$T = \frac{\theta J G}{L}$$
Eq. 23.4

The only shapes that this type of loading will be applied to is the circle and the circular tube. The value for J for these two shapes can be gotten from the routines *circle.m* and *obeam.m*. The value used is:

$$\text{Shaft: J} = \frac{\pi}{2}r^4$$

Eq. 23.5

$$\text{Tube: J} = \frac{\pi}{2}(R^4 - r^4)$$

This type of loading has a very fundamental difference from the previous axial and thermal loads. Torsional stress varies across the cross sectional area. As a general rule, engineers are concerned with the maximum stresses. These maximums occur on the outer edge where the radius is the largest possible. Stress states on the inner parts of the shaft could be calculated by changing the value of the radius of interest.

23.4   Output

The problems that are associated with this type of loading are solved very much like the previous types, axial and thermal.

Simplest situation: What are the reaction forces on the ends of the shaft as illustrated in Figure 23.1. The right-hand bracket has loosened and will not provide a restraining force until one degree of rotation has occurred:

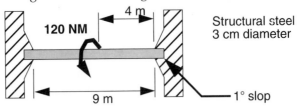

Figure 23.1   Simplest case of indeterminate torsional loading.

As always, gathering of the problem specific data comes first:

```
LengthRegions=[5 4]; %meters
TorqueLoad=120; %Newton meters
Gap=DR(1); %radians
G=matprop('structural steel','G','SI')*1e9;%Pascals
J=circle(0.03/2,'J'); %meters^4
Radius=0.015; %meters
```

The sign convention used here will be— ccw is a positive torque. On a beam that is viewed perpendicular to its axis, ccw will be defined as viewed from the right or the top.

Similar to the other statically indeterminate problems, one of the supports will be ignored to see what the deflection would be if unrestrained. Then the force required to bring the beam back to the desired position can be calculated. The length being torqued when the right-hand support is ignored is five meters. However when the force to counteract the calculated deflection is applied, the length will be the full nine meters:

```
Length=sum(LengthRegions); %meters
DeflectionFree=TorqueLoad*LengthRegions(1)/(J*G); %radians
Reactions(2)=-DeflectionFree*J*G/Length; %Newton meters
Reactions(1)=-Reactions(2)-TorqueLoad %Newton meters
Reactions =
  -53.3333  -66.6667
```

The stresses that occur in each region depend on the torque in that region, among other things. To find the torque acting on a region, take a cross section and sum the torque that must be acting there. In this simple case, it is the opposite of the *Reactions*:

```
TorqueRegions=-Reactions; %Newton meters
Stresses=TorqueRegions*Radius/J %Pascals
Stresses =
   1.0e+07 *
   1.0060  1.2575
```

This example could be expanded into more regions, with different cross sectional areas and more torques applied. An example of that nature will be covered in Section 29.4

Series torque loading: In this type of loading there is more than one force acting on the shaft. The shaft is rotating, but it is not accelerating so it is still in static equilibrium. Find the relative displacement of the different loading sections relative to the first. The bearing supports produce no moments. See Figure 23.2.

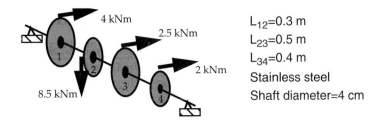

L$_{12}$=0.3 m
L$_{23}$=0.5 m
L$_{34}$=0.4 m
Stainless steel
Shaft diameter=4 cm

Figure 23.2   Quasi-static torsional loading.

Gathering the date into data vectors:

CH2302.m

```
LengthRegions=[0.3 0.5 0.4]; %meters
G=matprop('stainless steel','G','SI')*1e9;%Pascals
J=circle(0.02,'J');%meters^4
Radius=0.02; %meters
Area=circle(Radius, 'area');%meters^2
```

The torque that is acting on the first region, between torques one and two, can be found with a cross section drawn at any point in between the two torques. By inspection, the torque is 40 Nm. In the same manner the load in every section can be gathered:

TorqueRegions=[4 -4.5 -2]*1e3; %Newton meters

Now the displacement of each loading area can be calculated. Each of the displacements calculated will be in the form of "n+1 relative to n" or "two relative to one", "four relative to three."

RelativeDisplacements=TorqueRegions.*LengthRegions/(J*G);
RD(RelativeDisplacements)

```
ans =
    3.6476 -6.8392  -2.4317
```

To find all of the displacements relative to the first torque loading the relative displacements must be added. This can easily be done with the built-in command *cumsum*.

DisplacementsFromOne=cumsum(RelativeDisplacements);
RD(DisplacementsFromOne)

```
ans =
    3.6476 -3.1916  -5.6233
```

Because the torques in each region have already been calculated, it will be easy to find the stresses:

Stress=TorqueRegions*Radius/J

```
Stress =
   1.0e+08 *
    3.1831 -3.5810  -1.5915
```

In this example, it would have been easy to have the shaft change diameters or materials by changing the scalars G, J, and Area into vectors representing the values for each region rather than having a scalar that is valid in all regions.

23.5 Summary

Two different torque loading routines were covered in this section. The first was the statically indeterminate loading. This resembled the previous loading types in the method of solution. A redundant support is ignored, then the free deflection is calculated. The last step calculated the force that the ignored support must give to force the member back into shape.

The second problem is a quasi-static case where the relative displacements of the different torque application areas are calculated along with the stresses.

24

Shear, Moment, and Other Diagrams

24.1 Introduction

In previous sections, references have been made to loading diagrams for axial, torsional, shear and moment diagrams. All of these are useful for visualization of loading states on members. They make calculations at every point of the linear structure possible. Usually, these diagrams are only used for shearing and moments, and then used in bending problems. However, the mathematics that govern the creation of shear diagrams can be applied to create other types of diagrams.

24.2 Problem

Draw a diagram to represent the loading state of a beam or similar engineering structure. The types of data that should be applicable to this are:

- Area
- Axial loading
- Torsional loading
- Shear loading
- Moment loading
- Stresses

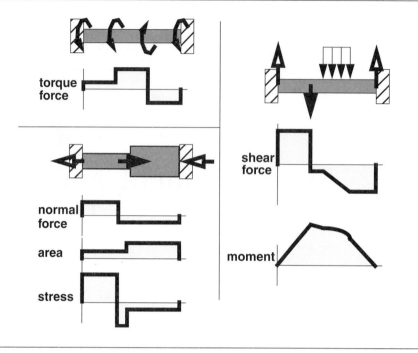

Figure 24.1   Different loading diagrams.

These diagrams, Figure 24.1, can be very useful not only for visualization tools, but also computationally. When the normal force diagram is "divided" by the area diagram, then the stress diagram is drawn. Notice that the area no longer needs to stay constant in a given region for the stress to be calculated.

24.3   Theory

The representation that will be used for these diagrams will be one of finite differences. A given beam will be divided along it's length into any given number of equally distributed points. The value of the measured quantity will be noted at each point along the beam. For example, if the following loading situation were to be represented with this method, the data points would appear as shown in Figure 24.2.

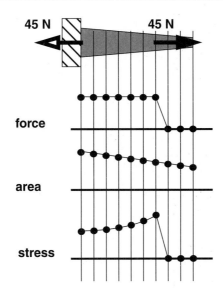

Figure 24.2   Graphical representation of loading.

The more data points that are taken, the better the results will be. It can be seen that at points where the force drops abruptly, the diagram falsely draws a short but steep slope. However, as the data points become more and more dense along the length of the beam, these transients become less and less noticeable.

The routines to be introduced are just expedient ways of creating data vectors to plot. The routines do very little mathematically. For example, the data vectors that would create the plots in Figure 24.2 are really simple as shown in Table 24.1.

	1	2	3	4	5	6	7	8	9	10
Force	45	45	45	45	45	45	0	0	0	0
Area	12.6	11.5	10.5	9.4	8.4	7.3	6.3	5.2	4.2	3.1
Stress	3.58	3.91	4.30	4.77	5.37	6.14	0	0	0	0

Table 24.1   Data used to generate Figure 24.2

There are three distributions that are common in this type of diagraming:
- Constant value: point load
- Linearly changing value: tapered area
- Cumulative value: distributed load

Constant value: If the diagram is to show a constant value over a range then this is the distribution to use, see Figure 24.3. A point load will cause a constant value until another load interrupts. Another use of this distribution is in the making of area diagrams.

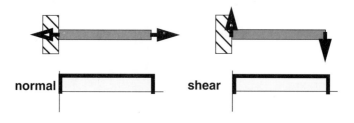

Figure 24.3   Constant value distribution.

Linearly changing value: If the diagram is to show a value that steadily changes from one amount to another then this is the distribution to use, see Figure 24.4. Most often this will be used for areas, not for loading. This is not the proper routine for distributed loading. The value does not necessarily have to change, it can simply be a constant value over a region.

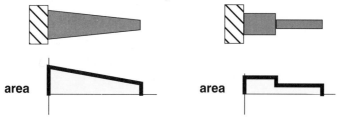

Figure 24.4   Linearly changing distribution.

Cumulative loading: This is the distribution to use for distributed loading, see Figure 24.5. With distributed loading, the value of the load is actually equal to the integral of the load from the beginning to the point of interest. This type of distribution accounts for this effect.

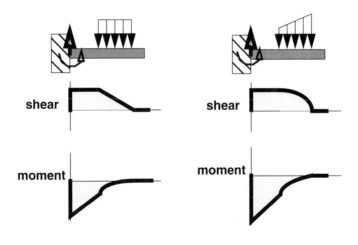

Figure 24.5   Cumulative loading distribution.

The cumulative loading routine finds the area under the loading curve at each point and applies it as a shear force. The area is always that of a trapezoid. In special cases, like the left-hand drawing in Figure 24.5, the trapezoid degenerates into a rectangle, and the area can always be found in that manner.

When the shear diagrams are used to create moment diagrams, the technique is to find the integral of the curve drawn in the shear diagram. The numerical method of integration that is used is the trapezoid method. Because of the use of this method, the more data points that are used, the better the results will be.

The routines to be introduced do not calculate anything so much as they format data into easily plotted vectors.

24.4 Template

```
function [y]=diagram (x,option,mag,place)
%DIAGRAM Creates vectors for use in plotting of diagrams.
%    DIAGRAM(X,OPTION,MAG,PLACE) will create a data vector the same length as
%    the input X.  The data will be created to one of three options:
%
%    Options:
%    'point' all data points that correspond to X>=PLACE are set equal to
%        MAG.  This is useful in shear diagrams when a point load is used.
%    'linear' all data points that correspond to PLACE(1)<=X<=PLACE(2) are
%        set equal to the linear interpolation of the values MAG(1) and
%        MAG(2).  This is useful when describing the area of a shaft that is
%        constant over a set length or that changes linearly over a set
%        length.  This is not for distributed loads.
%    'distributed' all data points are created to reflect the load from a
%        linearly distributed load that starts at PLACE(1) with a MAG(1) and
%        changes to MAG(2) at PLACE(2).
%
%    See also DIAGRAMINTEGRAL, DISPLACE.


%    Details are to be found in Mastering Mechanics I, Douglas W. Hull,
%    Prentice Hall, 1999


%    Douglas W. Hull, 1999
%    Copyright (c) 1999 by Prentice Hall
%    Version 1.00

option=lower(option);
y=zeros(size(x));

if strcmp(option,'point')
  VI=find(x>=place); %Valid Indices
  y(VI)=mag;

elseif strcmp(option,'linear')
  VI=find(x>=place(1) & x<=place(2)); %Valid Indices
  y(VI)=(x(VI)-place(1))/(place(2)-place(1))*(mag(2)-mag(1))+mag(1);

elseif strcmp(option,'distributed')
  VI=find(x>=place(1) & x<=place(2)); %Valid Indices
  BVI=find(x>place(2)); %Beyond Valid Indices
  height(VI)=(x(VI)-place(1))/(place(2)-place(1))*(mag(2)-
mag(1))+mag(1);
  y(VI)=(x(VI)-place(1)).*(height(VI)+mag(1))/2;
  y(BVI)=mean(mag)*(place(2)-place(1));
```

```
else
  disp ('Use a proper loading option: ''point'', ''linear'', or
''distributed''')
  return
end
```

diagramintegral.m

```
function [area]=diagramintegral(x,y)
%DIAGRAMINTEGRAL Integral of the given numerical data.
%   DIAGRAMINTEGRAL(X,Y) finds the integral of the data vector Y that
%   corresponds to the indices X.  The trapezoid method is used for
%   integration.  The routine is intended to work with the output from
%   DIAGRAM to create a moment diagram from a shear diagram, or other such
%   integrations.
%
%   See also DIAGRAM, DISPLACE.

%   Details are to be found in Mastering Mechanics I, Douglas W. Hull,
%   Prentice Hall, 1999

%   Douglas W. Hull, 1999
%   Copyright (c) 1999 by Prentice Hall
%   Version 1.00

if length(x)~=length(y)
  disp('Vectors must be the same length.')
  return
end

DeltaX=x(2)-x(1);
n=length(x);

subarea(1)=0;
for gapli=2:n %Generic All Purpose Looping Index
  subarea(gapli)=(y(gapli-1)+y(gapli))*DeltaX/2;
end

area=cumsum(subarea);
```

```
function []=plotSMD(x,shear,moment,displacement)
%PLOTSMD Plots a Shear Moment and optional Displacement diagram.
%    PLOTSMD(X,SHEAR,MOMENT,DISPLACEMENT) is a quick routine to show the
%    SHEAR, MOMENT and optional DISPLACEMENT diagrams on the same figure.
%    This routine can and should be modified to specific needs.
%
%    See also plotSMSD.


%    Details are to be found in Mastering Mechanics I, Douglas W. Hull,
%    Prentice Hall, 1999


%    Douglas W. Hull, 1999
%    Copyright (c) 1999 by Prentice Hall
%    Version 1.00

subplot(nargin-1,1,1)
plot ([0,x],[0,shear])
title ('Shear')
expandaxis; showx

subplot(nargin-1,1,2)
plot ([0,x],[0,moment])
title ('Moment')
expandaxis; showx;

if nargin==4
  subplot (nargin-1,1,3)
  plot (x,displacement)
  title ('Displacement')
  expandaxis; showx;
end
```

`interpolate.m`

```
function [outvalue]=interpolate(x,y,invalue)
%INTERPOLATE Linear interpolation for a given value.
%
%    INTERPOLATE(X,Y,XVALUE) With a vector of X and Y values that
correspond to
%    one another, the linear interpolation of the YVALUE that
corresponds to the
%    given XVALUE will be found.

%    Details are to be found in Mastering Mechanics I, Douglas W. Hull,
%    Prentice Hall, 1999

%    Douglas W. Hull, 1999
%    Copyright (c) 1999 by Prentice Hall
%    Version 1.00

if invalue>max(x) | invalue<min(x)
  disp('That value is not in the range that can be interpolated.')
end

biggerindex=min(find(x>=invalue));
smallerindex=max(find(x<=invalue));

X2=x(biggerindex);
Y2=y(biggerindex);

X1=x(smallerindex);
Y1=y(smallerindex);

if X2==X1
  outvalue=Y1;
else
  outvalue=((invalue-X1)/(X2-X1))*(Y2-Y1)+Y1;
end
```

```
>>x=0:0.1:10; %zero to ten by tenths
>>offset=3;
>>[F,P]=distload(-20,-40,3);
>>s(1,:)=diagram(x,'distributed',[-20 -40],[3 6]);
>>s(2,:)=diagram(x,'point',-F,0);
>>shear=sum(s);
>>area=diagram(x,'linear',[2 1],[0 10]);
>>stress=shear./area;
>>m(1,:)=diagram(x,'point',-F*(P+offset),0);
>>m(2,:)=diagramintegral(x,shear);
>>moment=sum(m);
>>figure(1)
>>plotSMD(x,shear,moment);
>>figure(2)
>>plot(x,stress)
>>title ('Stress')
```

24.5   Output

The data is not terribly complicated. Once the data vectors are gathered together, they can be plotted just as any other vectors in MATLAB. To construct shear-moment diagrams in MATLAB, starting from just the free body diagram.
1. Find all reaction forces and moments
2. Create a distance vector
3. Create a matrix of shear forces, one force per row
4. Sum the shear force matrix to create total shear vector
5. Create moment matrix, one moment per row
6. Make the last row in the moment matrix the integral of the total shear vector
7. Sum the moment matrix to create the total moment vector
8. Plot the total shear and the total moment vectors versus the distance vector

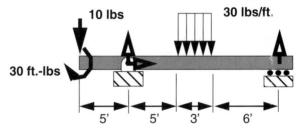

Figure 24.6   A beam in shear.

The above list will be followed in the context of solving the problem presented in Figure 24.6.
1. Following the methods presented in the statics section of this text, find the reaction forces and moments.

CH2401.m

```
MomentOnEnd=30;
[DLForce DLPlacement]=distload(-30,-30,3);
af(1,:)=[0 -10 0 0];
af(2,:)=[0 DLForce 10+DLPlacement 0];
Unknowns=[DR(90) 5 0; DR(90) 19 0;0 5 0];
Reactions=threevector(af,Unknowns,-MomentOnEnd);
```

2. Create a distance vector. Make the step size as small as computation time will reasonably allow. Make the step smaller after testing is done. Notice in this example that the moment does not equal zero as it should at the end of the beam. However, if the step size is made small enough, the error becomes insignificant.

```
x=0:0.1:20;
```

3. Each of the forces acting on the beam warrants its own row in the matrix. The notation *s(1,:)* is read as: "Matrix S row one and as many columns as is needed."

```
s(1,:)=diagram(x,'point',af(1,2),af(1,3));
s(2,:)=diagram(x,'point',Reactions(1,2),Reactions(1,3));
s(3,:)=diagram(x,'distributed',[-30 -30],[10 13]);
s(4,:)=diagram(x,'point',Reactions(2,2),Reactions(2,3));
```

4. Sum the shears that were gathered in the previous step into the total shear vector.

```
Shear=sum(s);
```

5. Create the moment matrix placing each moment in its own row in the matrix, be sure to use the proper sign convention.

```
m(1,:)=diagram(x,'point',MomentOnEnd,0);
```

6. The last row, which may also be the first if there are no applied moments, is created with the *diagramintegral* function.

```
m(2,:)=diagramintegral(x,Shear);
```

7. Sum the moments that were gathered in the previous two steps into the total moment vector.

```
Moment=sum(m);
```

8. Plot the shear and moment diagrams. A simple function to do this has been created. If a more customized shear and moment diagram is needed the plotting can be done with plotting commands of your specification.

```
plotSMD(x,Shear,Moment)
```

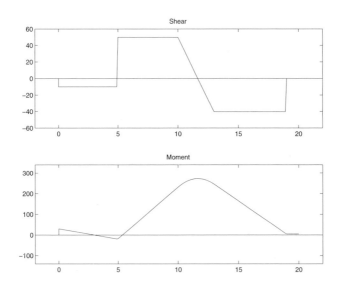

Figure 24.7   Shear and moment diagrams made by MATLAB.

This little amount of work has gotten the diagrams in Figure 24.7, but there is more that can be done with the data that has been gathered. If the data you want is at a set length and that specific length is not stated, like 12.33 when the vector is . . . 12.2, 12.3, 12.4, . . . then an interpolate function would be useful. For example, to find the value of the shear at the point $x=12.33$

```
>>interpolate (x,Shear,12.33)
ans =
 -20.2571
```

If the largest absolute value of shear or moment was to be found:

```
>>MaxShear=max(abs(Shear))
MaxShear =
   49.6429
>>MaxMoment=max(abs(Moment))
MaxMoment =
  272.2393
```

To find the x value at the largest shear or moment:

```
>>XatMaxMoment=x(find(abs(Moment)==max(abs(Moment))))
XatMaxMoment =
   11.7000
>>XatMaxShear=x(find(abs(Shear)==max(abs(Shear))));
```

The second command will return several values of x because the value plateaus for the range between 5.00 and 10.00.

24.6   Features

These routines are very powerful for solving an equation for an entire beam rather than for just one position. This makes graphing of values easier. Common applications of these routines would be to calculate stress, area, axial, torsional, and shear forces. In this way the maximum or minimum forces would be easy to find.

24.7   Summary

Required argument, optional argument, [**ssv**] = Stress State Vector

For *diagram.m*, *index* is a vector representing the many different distances along the beam. This vector is often constructed as **x=0:0.1:1;** In this way x would equal [0.0, 0.1, 0.2, 0.3,... 0.9, 1.0].

There are three different modes that *diagram.m* can be run in: point, linear, and distributed. Often this routine is used to simulate loading conditions. When doing that, each of the different loads is to be introduced separately, and then the total of the loads can be found using the sum command.

diagram(*index*,'point',*magnitude*,*distance*)

In the *'point'* mode, *diagram.m* will create a vector the same length as *index*. It will find all the values of *index* greater than or equal to the *distance* input and assign the value *magnitude* to them in the new matrix. This is to be used with point loads, or with point moments in conjunction with the *diagramintegral.m* routine.

diagram(*index*,'linear',[*magnitude*],[*distance*])

In the *'linear'* mode, *diagram.m* will create a vector the same length as *index*. It will find all the values of *index* that fall between the two values in the *distance* input vector. The routine will assign a value that is linearly interpolated from the two values in the *magnitude* vector based on the position between the two ends. This mode is useful for creating an area profile for a beam.

diagram(*index*,'distributed',[*magnitude*],[*distance*])

In the *'distributed'* mode, *diagram.m* will create a vector the same length as *index*. It will find all the values of *index* that fall between the two values in the *distance* input vector. The routine will assign a value based on the amount of area under the curve that lies to the left of the point of interest. Note: After the valid range of the distributed load, this routine will still find all the area that lies beneath the curve for the rest of the beam.

The best method to understanding the subtleties between these three modes is to use the routine a few times and graph the values.

diagramintegral(*x*,*y*)

The vectors *x* and *y* represent many points on a curve, usually a shear diagram curve. The output from this routine will be a vector with the value of the integral, or the area under the curve, at that point. This routine uses the trapezoid rule to evaluate the integral numerically.

plotSMD(*x*, *shear*, *moment*, *displacement*)

This routine will simplify the construction of shear, moment, and displacement diagrams. It can and should be customized to your personal needs. This represents a very generic and simple set of diagrams. Additional labeling and color changing may be desired. The optional displacement characteristic will be covered in Chapter 30.

interpolate(*x*,*y*,*x value*)

Given a set of *x* and *y* data that represents a function, and the desired *x value*, the routine will use linear interpolation to find the unknown *y* value.

25

Flexure Loading

25.1 Introduction

The next type of loading to be discussed is flexure, or bending loading. In many ways this is the most important of the loading types. As a general rule, engineers design against failure due to bending. This is because if an element is capable of withstanding the applied bending load, then it is likely more than strong enough to withstand the other forms of stress that exist.

25.2 Problem

When an element is subjected to a moment, possibly caused by a force acting at a distance, find the stresses imposed anywhere on the cross section of interest.

25.3 Theory

The equation that governs this type of loading is:

$$\sigma = \frac{My}{I}$$

Eq. 25.1

Where M is the moment at that point in the beam, I is the moment of area of the cross section about the neutral axis, and y is the distance from the neutral axis to the point of interest. Once the stress is found, it can be manipulated like any other form of stress that is invoked on a member.

The sign convention with the y value is negative above the neutral axis, and positive below. This switch in convention makes the stresses negative for compression.

Notice that this type of stress varies across the cross section. Engineers are usually interested in the highest stresses, those will be found at the point that is farthest from the neutral axis. Notice also how quickly the moment can grow very large simply by increasing the moment arm. This is what causes bending moments to be the limiting stresses.

25.4   Output

These problems rely very much on the use of the moment diagrams that were developed in the Chapter 24. Their use allows the bending stress to be found throughout the beam in one step. With the other types of loading, the beam could usually be broken into several finite regions that could be easily solved. However, with bending, the beam is treated as a continuous entity and so cannot be broken into regions as the other types of loading allowed.

Simplest situation: What is the stress on the beam at the indicated point in Figure 25.1 due to the bending moment?

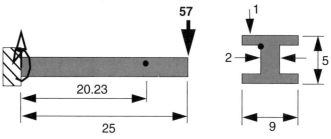

Figure 25.1   Stresses due to bending.

This problem will be done in a general manner:

CH2501.m

```
BeamLength=25;
PointLength=20.23;
AppForce=-57;
CSBase=9;
CSHeight=5;
CSWeb=2;
CSThickness=1;
YPosition=4;
```

Putting all these values in parametrically will allow the values to be easily changed. This is especially important as the problems become more complex.

Next, find the reaction forces and the shear and moment diagrams. Remember that the sign convention for moments switches when dealing with shear moment diagrams:

```
[WallForce WallMoment]=reaction([0 AppForce BeamLength 0],[0,0]);
x=[0:.01:BeamLength];
s(1,:)=diagram(x,'point',WallForce(2),0);
s(2,:)=diagram(x,'point',AppForce,BeamLength);
Shear=sum(s);
m(1,:)=diagram(x,'point',-WallMoment,0); %switch sign convention
m(2,:)=diagramintegral(x,Shear);
Moment=sum(m);figure(1)
plotSMD(x,Shear,Moment);
```

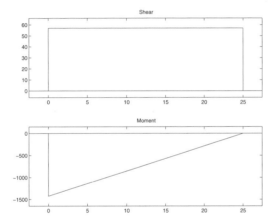

Next is to deal with the cross sectional information:

```
CSIx=ibeam(CSBase,CSHeight,CSThickness,CSWeb,'I','Ix');
CSNA=ibeam(CSBase,CSHeight,CSThickness,CSWeb,'I','centY');
YValue=CSNA-YPosition;
```

Finally, the stress can be calculated:

```
MomentValue=interpolate(x,Moment,PointLength);
StressValue=MomentValue*YValue/CSIx
StressValue =
    5.2287
```

This gets the desired value. If the point of interest changed, it would be easy enough to change the variables *PointLength* and *YPosition* to reflect this change. However, changing these values for several different points along the beam would become tedious. To do them all at once in graphical form would be similar to creating a share moment diagram. This will be a stress diagram:

```
>>StressVector=Moment*YValue/CSIx;
>>figure(2)
>>plot(x,StressVector)
>>title ('Stress')
```

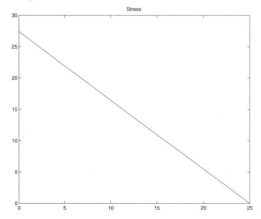

This does not happen to be the most aesthetically pleasing of stress diagrams, but it is not that interesting of a problem either. More complicated problems will lead to more interesting diagrams. This diagram is consistent with the single point that was gathered earlier.

More complicated bending stresses: What is the maximum stress in the bar due to bending? Where is this maximum stress due to bending located? See Figure 25.2.

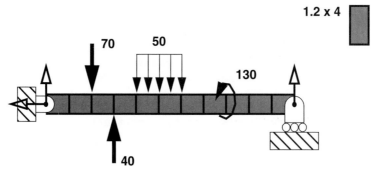

Figure 25.2   Stresses due to bending

First the problem data will be brought into variables:

CH2502.m

```
fmag=[-70 40];
fplace=[2 3];
dismag=[-50 -50];
disstart=4;
dislength=2;
beamlength=11;
couplemag=130;
coupleplace=8;
disend=disstart+dislength;
cswidth=1.2;
csheight=4;
Yposition=4;
```

Next the reaction forces will be solved for using the methods of statics:

```
af=[[0 0]' fmag' fplace' [0 0]'];
[DisForce, DisPlace]=distload(dismag(1),dismag(2),dislength);
af(3,:)=[0 DisForce 4+DisPlace 0];
UnknownPlacement=[DR(90) 0 0;DR(90) beamlength 0;DR(180) 0 0];
Resultants=threevector(af,UnknownPlacement,couplemag);
```

The shear and moment diagrams will be calculated and drawn:

```
x=0:0.01:11;
s(1,:)=diagram(x,'point',Resultants(1,2),Resultants(1,3));
s(2,:)=diagram(x,'point',fmag(1),fplace(1));
s(3,:)=diagram(x,'point',fmag(2),fplace(2));
s(4,:)=diagram(x,'distributed',dismag,[disstart disend]);
s(5,:)=diagram(x,'point',Resultants(2,2),Resultants(2,3));
Shear=sum(s);
%%%
m(1,:)=diagram(x,'point',-couplemag,coupleplace);
m(2,:)=diagramintegral(x,Shear);
Moment=sum(m);
%%%
figure(1)
plotSMD(x,Shear,Moment);
```

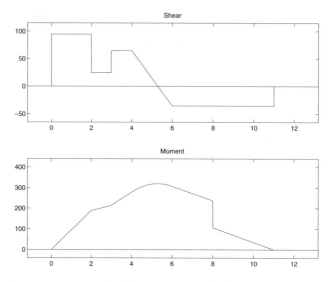

Next the cross sectional information is calculated and plotted:

```
CSIx=rectangle(cswidth,csheight,'Ix');
CScentY=rectangle(cswidth,csheight,'centY');
YValue=CScentY-Yposition;
Bending=Moment*YValue/CSIx;
figure(2)
plot (x,Bending)
title ('Bending Stress')
```

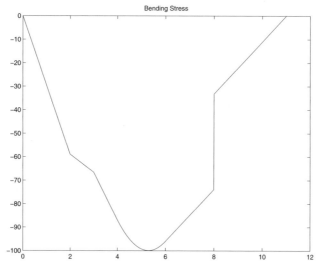

Finally, the maximum value and its position are pulled out:

MaxBend=max(abs(Bending))

MaxBend =
   99.9040

MaxBendAt=x(find(abs(Bending)==MaxBend))

MaxBendAt =
   5.2900

It may be interesting to run the M-file again changing the initial conditions to see how the positioning of forces and moments affects the results.

25.5   Summary

Finding the stress state at a point due to bending is an extension of the data gathered for shear and moment diagrams. Bending is the most important of loading types because it is the method of failure that is most often the limiting factor in a design.

26

Transverse Shear Loading

26.1 Introduction

In earlier sections, the average shear on a cross section was calculated. This is a useful number to know, but it is not entirely accurate. The transverse shear, the shear acting on a cross section, is not uniformly distributed across the face. This means that the shear at the upper edge of a beam is different from the shear stress that is acting on the centroid of the beam. This distribution is important when trying to calculate the stress acting upon a differential element of the beam.

26.2 Problem

What is the shear stress on a differential element located arbitrarily on the cross section? See Figure 26.1.

Figure 26.1   Shear stress distribution.

26.3 Theory

The governing equation in transverse shear is:

$$\tau = \frac{VQ}{IT}$$

Eq. 26.1

Where:

- τ is shear stress
- Q is $\bar{y}A$
- I is the second moment of inertia about the axis perpendicular to the force
- t is the thickness at the point of interest

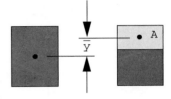

• centroid

Note that either the area above, or below, the point of interest may be used for calculations. Use whichever will yield the simpler calculations.

There are some assumptions that go into this theory that cause little difficulty, but should be recognized. The first assumption is that the shear stress across the width of an object is constant. This is not true of course, but it is a reasonable assumption. The second problem is at flange web junctions. These actually act like stress concentrators. Also at these points the inner regions of the flanges are free edges, so they should have zero stress. These difficulties have little bearing on practical engineering practice.

26.4 Output

Shear stress on a simple shape: What is the shear force acting on each of the labeled differential elements illustrated in Figure 26.2?

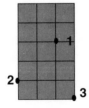

Shear force of 10 N is acting straight down on the cross section

Squares are 1 cm

Figure 26.2   Cross section in shear.

This problem will be done once, and then the initial values can be changed for each of the other points by changing the trial number:

```
trial=1
YValue=[3,1,0];
Shear=-10;
CSWidth=0.03; %meters
CSHeight=0.05; %meters
I=rectangle(CSWidth,CSHeight,'Ix');
Thickness=CSWidth;
QArea=rectangle(CSWidth,YValue(trial),'area');
WholeCentroid=rectangle(CSWidth, CSHeight, 'centY');
PartCentroid=rectangle(CSWidth, YValue(trial), 'centY');
QYbar=abs(WholeCentroid-PartCentroid);
Q=QArea.*QYbar;
ShearAtPoint=Shear*Q/(I*Thickness);
```

Running this program three times, each time changing the trial number, will lead to the three different answers shown in Table 26.1.

Trial	Y value (cm)	Shear force (Pascals)
1	3	-141600000
2	1	-15200000
3	0	0

Table 26.1   Answers to shear problem.

This example was easy, because the code did not have to be altered for each of the runs. However, in more complicated shapes it will become necessary to do so.

Shear stress on complex shapes: What is the shear stress acting at each of the labeled points? See Figure 26.3

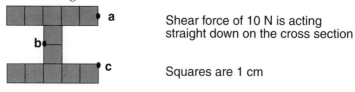

Shear force of 10 N is acting straight down on the cross section

Squares are 1 cm

Figure 26.3   I-Beam in shear.

With this problem, see Figure 26.4, each of the three points will have to be dealt with in a slightly different manner:

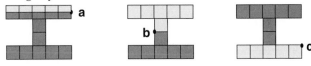

Figure 26.4   Different methods for finding Q.

Each of the three elements has to calculate Q in a different manner. In the first and last, the area to be found is a rectangle, but the rectangle is located in different places. In the second, the area is a T-beam cross section. In the third example, the choice for the thickness value is questionable. It could be one or five.

For these reasons, the code will have to be written for each of the different situations:

CH2602.m

```
Shear=-10; %Newtons
AYValue=0.035; %meters
BYValue=0.02; %meters
CYValue=0.01; %meters
CSW=0.05; %meters
CSH=0.04; %meters
CSBase=0.01; %meters
CSWeb=0.01; %meters
I=ibeam(CSW,CSH,CSBase,CSWeb,'I','Ix');
AThickness=CSW;
BThickness=CSWeb;
CThickness=CSWeb;
```

Notice that the dilemma of what value to choose for thickness was solved by choosing the smaller of the two. This will yield the more conservative value for stress.

```
OldCentroid=ibeam(CSW,CSH,CSBase,CSWeb,'I','centY');
AQArea=rectangle(CSW,CSH-AYValue,'area');
AQNewCen=rectangle(CSW,CSH-AYValue,'centY')+AYValue;
AQYBar=abs(OldCentroid-AQNewCen);
AQ=AQYBar*AQArea;
```

Now the same procedure for the other two elements:

```
BQArea=tbeam(CSW,CSH-BYValue,CSBase,CSWeb,'n','area');
BQNewCen=tbeam(CSW,CSH-BYValue,CSBase,CSWeb,'n','area')+BYValue;
BQYBar=abs(OldCentroid-BQNewCen);
BQ=BQYBar*BQArea;
CQArea=rectangle(CSW,CSH-CYValue,'area');
CQNewCen=rectangle(CSW,CSH-CYValue,'centY');
CQYBar=abs(OldCentroid-CQNewCen);
CQ=CQYBar*CQArea;
```

Finally, the stresses may be calculated:

```
AStress=Shear*AQ/(I*AThickness);
BStress=Shear*BQ/(I*BThickness);
CStress=Shear*CQ/(I*CThickness);
```

Trial	Y value (cm)	Shear force (Pascals)
A	3.5	-3645.8
B	2	-33333.0
C	1	-31250.0

Table 26.2   Answers to shear problem.

These stresses, as shown in Table 26.1, look reasonable since the stress increases as the point of interest gets farther from the centroid. Also the stress decreases by an order of magnitude when the region of the wide flange is entered.

26.5   Features

Computationally this is the most strenuous of the loading types. This is because of the requirement for finding the Q value and the thickness. The Q value is difficult because it requires the finding of an area and a centroid. Because all of these values depend heavily on the placement of the element within the cross section, it is very difficult to write routines to calculate the values automatically for complex cross sections like I-beams or channels.

Even though there are a few more steps in solving this type of stress, there are routines available to make the job easier. Solving this type of problem in MATLAB is still faster than doing it with pencil and paper.

26.6   Summary

The formula for the transverse shear is straightforward. The most difficult part is finding the value for Q. This value is found in actually two steps. The first is to find the area above, or below, the point of interest. The next step is to find the distance between the centroid of the area found in the first step and the centroid of the whole cross section.

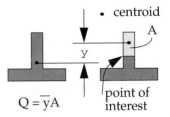

$$Q = \bar{y}A$$

As a check of the answers that are given by MATLAB, a few simple rules should be remembered. First, the stress decreases as the point of interest moves away from the centroid of the cross section. Second, the stress at the point of interest decreases as the cross section width increases at the point of interest.

27

Pressure Loading

27.1   Introduction

Thin-walled pressure vessels are common in industry. A pressure vessel may be considered thin walled if the radius of the vessel is more than ten times the thickness of the vessel. Common vessels are tanks and boilers. The stresses in a thin-walled vessel do not vary significantly and may be considered constant. The pressure in the vessel is considered to be positive gauge pressure.

27.2   Problem

What are the stresses acting on a differential element in a thin walled pressure vessel as shown in Figure 27.1?

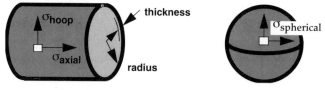

Figure 27.1   Notation of thin- walled pressure vessels.

27.3   Theory

The governing equations in this type of loading are:

$$\sigma_{spherical} = \sigma_{axial} = \frac{\text{pressure} \times \text{radius}}{2 \times \text{thickness}}$$

$$\sigma_{hoop} = \frac{\text{pressure} \times \text{radius}}{\text{thickness}}$$

Eq. 27.1

Notice that the spherical and axial stresses are the same, and that they are both double the hoop stress. Some pressure vessels are capped with spherical ends. Stress elements near these rounded ends should be considered to be in spherical stress since it is the larger of the two.

27.4 Output

What is the hoop stress in a cylindrical vessel if it has the following characteristics? See Figure 27.2.
- Thickness: 0.1 inches
- Inner diameter: 8 inches
- Pressure: 50 psi

As always, first the relevant values are gathered together:

CH2701.m

```
Thickness=0.1;
ID=8;
Pressure=50;
```

The equation is entered and the answer is given:

```
HoopStress=Pressure*(ID/2)/Thickness
HoopStress =
    2000
```

What is the stress state of the element represented? What is the maximum shearing stress and the primary stresses? Assume plane stress.

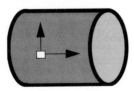

pressure: 400 kPa
thickness: 25 mm
radius: 220 mm

Figure 27.2   Thin-walled pressure vessel.

Gathering the data:

```
Pressure=400;
Thickness=25;
Radius=220;
```

Calculating the hoop and axial stresses:

StressHoop=Pressure*Radius/Thickness

StressHoop =
    3520

StressAxial=Pressure*Radius/(2*Thickness)

StressAxial =
    1760

Setting up the stress state for Mohr's circle routine:

StressState=[StressAxial StressHoop 0];
mohrs (StressState, 'plane stress')

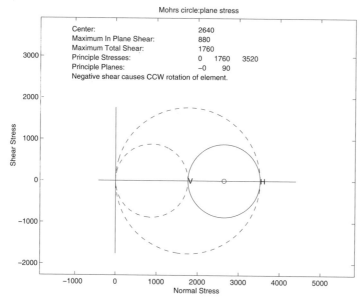

From this diagram the primary stresses and the maximum stress can be easily found.

This example is actually a lead into the combined loading covered in the next section. With little trouble, two different stress types, hoop and axial, were combined on a single element to find the total stress state. The same type of thinking can be applied to multiple stress types.

27.5   Summary

This type of problem is not very difficult to solve mathematically. It is included here mostly for completeness so that when combined loading is introduced, all the different types of loading may be represented.

28

Combined Loading

28.1   Introduction

The previous chapters outlined the major loading types: axial, thermal, torsional, bending, shear, and pressure vessel. All of these loadings may exist at the same point in an object. Due to the principle of linear superposition, the effects of each of these stresses on an element may be added together to find the total stress on an element. Knowing the total stress state on the element allows for the manipulation of the stress.

28.2   Problem

Given an object that is subjected to multiple loading types, see Figure 28.1, what is the total stress state that is induced?

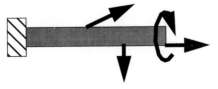

Figure 28.1   Combined loading state.

28.3   Theory

The principle of superposition allows the stress caused by each load within a loading type to be added together: The stresses caused by two axial loads may be added together. In the same way, if an element is stressed by an axial load and a bending load, the stresses may be combined to find the total stress state.

 The general procedure for these combined loading problems is to find the stresses caused by each loading type on the element of interest. When all the induced stresses are found, they may be added together to find the total stress state.

28.4   Output

This is best shown by way of examples, there are an infinite number of combinations of loading types available. However, once the techniques have been shown on a few examples, they can be applied to any situation as needed.

Bending and axial loading: In axial loading the force had to be applied to the centroid of the cross section. If it was not, then it would cause a bending moment. The theory behind axial loading did not cover the moment that would be caused by noncentric loading. However, the theory of bending loads covers noncentric loading, therefore the entire problem can be solved.

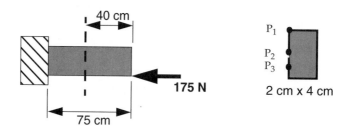

Figure 28.2   Bending and axial loading.

 In this method, the reaction forces that must be operating at the cross section are calculated and then the stresses figured from there, see Figure 28.2.

```
CSWidth=0.02; %meters
CSHeight=0.04; %meters
BeamLength=0.75; %meters
CSPosition=0.35; %meters
Force=[-175 0 BeamLength 0];%Newtons
CSCentY=rectangle(CSWidth,CSHeight,'centY'); %meters
CSArea=rectangle(CSWidth,CSHeight,'area'); %meters^2
CSIx=rectangle(CSWidth,CSHeight,'Ix'); %meters^4
[ReactForce ReactMoment]=reaction(Force,[CSPosition CSCentY]);
```
This gives the reaction force and moment that must be acting through the centroid of the cross section of interest. It is important that the force and moment are found at the centroid of the cross section. Knowing these forces, the axial stress can be found. Remember to use the proper convention so that compression is negative:
```
AxialStress=-mag(ReactForce,'x')/CSArea;
```
This stress is constant across the cross section, so the calculation need only be done once. However, the bending stress is going to be different at each of the three points of interest.
```
YPosition=[0.04 0.02 0.01]; %meters
YValue=CSCentY-YPosition; %meters
BendStress=-ReactMoment*YValue/CSIx;
```
How do these two stresses interact? Looking at a differential element as shown in Figure 28.3

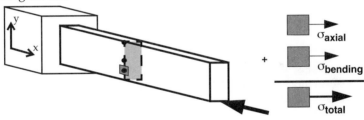

Figure 28.3   Combined loading.

These two stresses both happen to cause stresses that act upon the same plane. They both act on the vertical face of the stress elements causing a σ_x. These two stresses may simply be added together:
```
TotalStress=BendStress+AxialStress
TotalStress =
   437500   -218750   -546875
```

This example shows how the stress can vary over the cross section. Notice that the stress at the first position is 437 kN in tension, but the stress at the third position is actually 547 kN in compression. Notice that the second point is located on the neutral axis, for this reason the bending stress has no effect and the total stress at that point is actually equal to the axial stress alone.

If the stress were to be found at another point by changing the choice of cross sections, it could be easily done by changing the placement of the cross section.

Axial and pressure vessel stresses: In this example, a cylindrical barrel of the following dimensions, see Figure 28.4, is pressurized. This barrel also supports the weight of a second barrel that is stacked on top of it. What is the stress state at the indicated point on the bottom barrel? Draw a Mohr's circle representation of the stress state, as shown in Figure 28.5. Assume plane stress.

Inner diameter: 55 cm
Thickness: 0.8 cm
Height: 90 cm
Weight: 55 kg
Pressure 450 kPa

$\sigma_{axial} + \sigma_{weight}$

σ_{hoop}

Figure 28.4   Axial and pressure loading.

Gathering the data:

CH2802.m

```
ID=0.55; %meters
Thickness=0.002; %meters
Height=0.90; %meters
Mass=68; %kilograms
Pressure=4500; %Pascals
Gravity=-9.81; % meters/s^2
```

First, the axial stress due to the weight of the top barrel can be calculated:

```
Area=obeam(ID+2*Thickness,ID,'area');
Weight=Gravity*Mass;
StressNormal=Weight/Area;
```

Next, the stresses due to the pressure can be calculated:

```
StressAxial=Pressure*(ID/2)/(2*Thickness);
StressHoop=Pressure*(ID/2)/Thickness;
```

Now these three stresses can be gathered into a stress state for the element:

```
StressState=[StressHoop, StressAxial+StressNormal 0];
mohrs (StressState, 'plane stress')
```

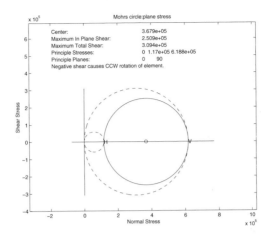

x 10⁵
Mohrs circle:plane stress

Center: 3.679e+05
Maximum In Plane Shear: 2.509e+05
Maximum Total Shear: 3.094e+05
Principle Stresses: 0 1.17e+05 6.188e+05
Principle Planes: 0   90
Negative shear causes CCW rotation of element.

Figure 28.5   Mohr's circle representation of stress state.

This example shows how the stress caused by the weight of the second barrel actually acts to lower the total stress on the element by causing some compression to counteract the tension caused by the pressure.

Shear, torsion, normal, and bending stresses: What is the stress state on the pictured element in Figure 28.6?

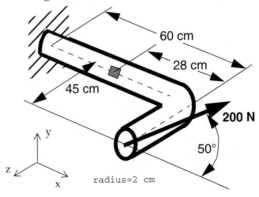

Figure 28.6   Many different types of loading combined.

After gathering all of the data into the proper variables, the next thing to do is to change this three dimensional problem into a two dimensional problem. The best way to do this is to change the given force from a force at a distance away from the x-y plane to a force-couple combination. This will provide a torque about the x axis and a bending moment along the z axis. The plan is shown in Figure 28.7.

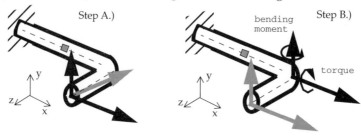

Figure 28.7   Moving forces to create a planar problem.

The MATLAB code to do all of this:

CH2803.m

```
BeamLength=0.60; %meters
ArmLength=0.45; %meters
ElementLength=0.32; %meters
Radius=0.02; %meters
Force=deg2xy([30,200,BeamLength,0]);
BendingMomentY=mag(Force,'x')*ArmLength;
NAZ=Radius;
BendingMomentZ=mag(Force,'y')*(BeamLength-ElementLength);
NAY=0;
Torque=mag(Force,'y')*ArmLength;
```

The problem has been neatly reduced to a planar problem. The next step is to find each of the stresses, Figure 28.8, that are induced:

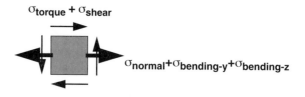

Figure 28.8   Stress element.

```
Area=circle(Radius,'area');
I=circle(Radius,'Iy');
J=circle(Radius,'J');
QArea=halfcircle(Radius,'n','area');
QYBar=halfcircle(Radius,'n','centY');
Q=QArea*QYBar;
Thickness=2*Radius;
StressNormal=mag(Force,'x')/Area;
StressBendingY=BendingMomentY*NAZ/I;
StressBendingZ=BendingMomentZ*NAY/I;
StressTorque=Torque*Radius/J;
StressShear=mag(Force,'y')*Q/(I*Thickness);
StressX=StressNormal+StressBendingY+StressBendingZ;
ShearXY=StressTorque+StressShear;
StressState=[StressX, 0, ShearXY];
mohrs (StressState, 'plane stress')
```

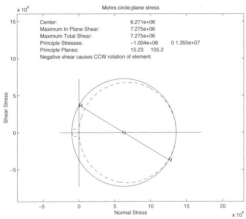

Figure 28.9   Mohr's circle representation.

28.5   Summary

This chapter links together the six different loading types that have been covered: thermal, normal, torsional, transverse shear, flexure, and thin-walled pressure vessel. Each of these types of stresses can be calculated independently and then summed together to find the complete stress state of a differential element. Once the stress state is found, it can be manipulated as explained in earlier chapters. The most common thing to do with a stress state, once found, is to make a graphical representation of it using the Mohr's circle, see Figure 28.9.

29
Loading — Examples

29.1 Series axial loading

All of the following problems, see Figure 29.1, are mathematically similar. They can all be solved with the same bit of code just by changing the initial conditions:

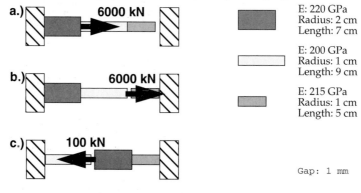

Figure 29.1   Different problems all solved with same code.

When entering the *Force* vector, forces that tend to close the gap are positive, those that tend to widen the gap are negative. In the *AppliedLoads* vector use the sign convention of positive to the right.

variation *a:* CH2901.m

```
Force=[6000 0 0]*1e3;%Newtons
Length=[0.07 0.09 0.05];%meters
Radii=[0.02 0.01 0.01];%meters
Youngs=[220 200 215]*1e9;%Pascals
Gap=0.001;%meters
AppliedLoads=[6000]*1e3;%Newtons
%%%Begin Common Code%%%%
Area=(Radii.^2)*pi;%meters^2
Deflection=(Force.*Length)./(Area.*Youngs);
TotalDeflection=sum(Deflection);
ForcedDeflection=TotalDeflection-Gap;
Stiffness=Length./(Area.*Youngs);
Percentage=Stiffness./sum(Stiffness);
RegionalDeflection=ForcedDeflection.*Percentage;
ReactionRightVector=-RegionalDeflection.*Youngs.*Area./Length;
RightReaction=ReactionRightVector(1)
LeftReaction=-sum(AppliedLoads)-RightReaction
```

The answers for variation a.): -5,786 kN and -214 kN for the left and right reactions. To solve this problem for the other two variations, the first six lines must be changed to reflect the relevant data:

variation *b:* CH2902.m

```
Force=[6000 6000 0]*1e3;%Newtons
Length=[0.07 0.09 0.05];%meters
Radii=[0.02 0.01 0.01];%meters
Youngs=[220 200 215]*1e9;%Pascals
Gap=0.001;%meters
AppliedLoads=[6000]*1e3;%Newtons
```

Variation b.) yields -2,243 kN and -3,757 kN for the left and right reactions.

variation *c:* CH2903.m

```
Force=[0 100 100]*1e3;%Newtons
Length=[0.09 0.07 0.05];%meters
Radii=[0.01 0.02 0.01];%meters
Youngs=[200 220 215]*1e9;%Pascals
Gap=0.001;%meters
AppliedLoads=[-100]*1e3;%Newtons
```

Variation *c* yields -271 kN and 371 kN for the left and right reactions. These answers do not pass the logical test. If these really were the reaction forces, the summation of forces would indeed equal zero, but looking at the free body diagram, see Figure 29.2, this is not a valid answer:

Figure 29.2   Variation c answers: assumptions incorrect.

These answers imply that a tension force is being supported across the 1 mm gap. This, of course, is impossible. As with all the routines in this book, it must be checked that any and all assumptions are true. The implied assumption of this problem is that the gap actually does close. The answers do not make sense in this final problem because the forces are not great enough to close the gap. In that case, it is a statically determinant problem and can be solved by inspection.

29.2   Parallel axial loading, known displacement

Although the following two problems look dissimilar, as shown in Figures 29.3 and 29.4, they can be solved with the same code by simply changing the initial conditions. For each of them, find the force carried by each member to achieve the pictured displacement.

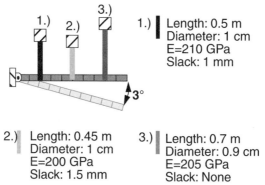

Figure 29.3   Cables stressed by a rigid bar: known displacement, CH2904.m.

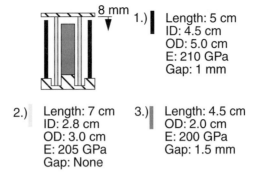

1.) Length: 5 cm
    ID: 4.5 cm
    OD: 5.0 cm
    E: 210 GPa
    Gap: 1 mm

2.) Length: 7 cm
    ID: 2.8 cm
    OD: 3.0 cm
    E: 205 GPa
    Gap: None

3.) Length: 4.5 cm
    OD: 2.0 cm
    E: 200 GPa
    Gap: 1.5 mm

Figure 29.4   Tubes under compression: known displacement, CH2905.m.

The situations are the same because several members are each carrying a certain part of the load caused by a known displacement. In the first problem, the displacement will be different for each of the different cables. For the second problem, the displacements will all be the same save for the free deflection given by the gaps. The second case should be thought of as a special case of the first, that is, all of the displacements could be different, but they happen to be the same.

Each of these problems will be set up separately, and then the code common to each of them will be presented. The setup must find all of the variables to satisfy Equation 21.5. These variables are commented as %NEEDED:

CH2904.m

```
Angle=DR(0.2);
Slack=[0.001 0.0015 0]; %meters
PicturedLength=[0.5 0.45 0.7]; %meters
Length=PicturedLength+Slack;%meters %NEEDED
WirePlacement=[0.2 0.5 0.8]; %meters
Horizontal=WirePlacement-cos(Angle)*WirePlacement;
Vertical=PicturedLength+sin(Angle)*PicturedLength;
NewLength=hyp(Horizontal,Vertical);
DeltaL=NewLength-Length; %meters %NEEDED
Radii=[0.01 0.01 0.009]/2;
Area=circle(Radii,'area');%meters^2 %NEEDED
E=[210 200 205]*1e9;%Pascals %NEEDED
```

```
Length=[0.05 0.07 0.045]; %meters %NEEDED
Displacement=0.008; %meters
Gap=[0.001 0 0.0015]; %meters
DeltaL=Displacement-Gap; %meters %NEEDED
ID=[0.045 0.028 0];
OD=[0.05 0.03 0.02];
ODArea=circle(OD,'area');
IDArea=circle(ID,'area');
Area=ODArea-IDArea; %meters^2 %NEEDED
E=[210 205 200]*1e9;%Pascals %NEEDED
```

Now that the two problems have been brought to same point, they can be completed with the same code:

Common code Example 29.2

```
Forces=DeltaL.*E.*Area./Length
```

First problem:
```
Forces =
1.0e+07 *
   4.3872    0.8538    3.6303
```
Second problem:
```
Forces =
1.0e+04 *
   2.4537    0.2463    4.5523
```
In this example, setting up the problem was most of the work. There was a common, but short, execution after the initial setup.

29.3   Thermal loading

Look back to Figure 22.4. Imagine that instead of being snug at the beginning of the problem, the sleeve is only 19.9 cm long, so there is a gap. How would this change the problem, and how should the code be modified?

The code should be easy to modify. Here is the corrected version, the explanation follows.

```
DeltaT=40;
LengthStartBolt=0.20; %meters
LengthStartSleave=0.1899; %meters
AreaSleave=obeam(0.09, 0.05, 'area'); %meters^2
AreaBolt=circle(0.02, 'area');%meters^2
AlphaBolt=matprop('bronze', 'thermal expansion', 'SI');%1/C
AlphaSleave=matprop('aluminum', 'thermal expansion', 'SI');%1/C
EBolt=matprop('bronze', 'E', 'SI')*1e9;%Pascals
ESleave=matprop('aluminum', 'E', 'SI');%Pascals
StrainFreeBolt=DeltaT*AlphaBolt;%unitless
StrainFreeSleave=DeltaT*AlphaSleave;%unitless
LengthFreeBolt=(1+StrainFreeBolt)*LengthStartBolt;%meters
LengthFreeSleave=(1+StrainFreeSleave)*LengthStartSleave;%meters
DeltaLengths=LengthFreeSleave-LengthFreeBolt;%meters
KBolt=LengthFreeBolt/(AreaBolt*EBolt);%meters/Newtons
KSleave=LengthFreeSleave/(AreaSleave*ESleave);%meters/Newton
DeltaBolt=DeltaLengths*KBolt/(KBolt+KSleave);%meters
DeltaSleave=-DeltaLengths*KSleave/(KBolt+KSleave);%meters
FinalLength=LengthFreeBolt+DeltaBolt
ForceBolt=DeltaBolt*EBolt*AreaBolt/LengthFreeBolt %Newtons
ForceSleave=DeltaSleave*ESleave*AreaSleave/LengthFreeSleave %N
StressBolt=ForceBolt/AreaBolt % Pascals
StressSleave=ForceSleave/AreaSleave %Pascals
```

In the original problem, the two initial lengths were the same. For this reason they were stored in a single variable. Because this variation has two different values for length, the variable *LengthStart* has been replaced by the two variables *LengthStartBolt, LengthStartSleave.*

The answer arrived at for the *FinalLength* has changed very little between the two problems: Resulting in 0.20013600000012 for the no gap problem and 0.20013599997369 for this new problem with a gap. The final length decreased as compared to the original problem. That seems logical since a lesser force was developed between the two members due to the gap.

This example shows how easily a program may be modified to reflect a new situation.

29.4 Torsional loading

Look back at the torsional loading problem in Figure 23.1. Imagine this problem being complicated by adding more loads, what would the support reactions be then?

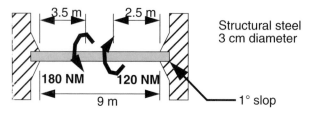

Structural steel
3 cm diameter

180 NM   120 NM

9 m

1° slop

Like the previous example, the code should be easy to modify to reflect the new situation:

CH2907.m

```
LengthRegions=[3.5 3 2.5]; %meters
AppliedTorques=[180 -120];
TorqueLoad=cumsum([-sum(AppliedTorques) AppliedTorques]);%Newton
meters
Gap=DR(1); %radians
G=matprop('structural steel','G','SI')*1e9;%Pascals
J=circle(0.03/2,'J'); %meters^4
Radius=0.015;
Length=sum(LengthRegions);
DeflectionFree=sum(TorqueLoad.*LengthRegions/(J*G)); %radians
Reactions(2)=-DeflectionFree*J*G/Length; %Newton meters
Reactions(1)=-Reactions(2)-sum(TorqueLoad) %Newton meters
TorqueRegions=cumsum([Reactions(1) AppliedTorques]); %Newton meter
Stresses=TorqueRegions*Radius/J %Pascals
```

This example only added a new torque to the original problem. The code is now equipped to handle any number of torque loads along the shaft.

```
Reactions =
 -43.3333 -16.6667
Stresses =
1.0e+07 *
  -0.8174    2.5779    0.3144
```

29.5 Bending

Since enough shear moment diagrams have been developed as parts of other examples there is no need to cover them again in this section. The only complication that can be added to bending flexure problems is to have a moment about an axis other than the x or y axes:

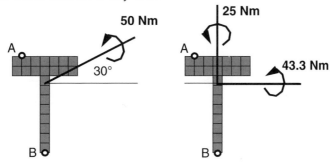

Figure 29.5   Breaking an eccentric moment into two orthogonal moments.

These two moments, Figure 29.5, can be dealt with independently. Find the stress on both the bottom and top faces.

CH2908.m

```
Base=0.07; %meters
Height=0.1; %meters
BaseHeight=0.02; %meters
WebT=0.01; %meters
PointX=0.01; %placement of point A
PointY=0.1; %placement of point A
%PointX=0.035; %placement of point B
%PointY=0; %placement of point B
Moment=50; %Newton meters
Angle=DR(30);
XAxisMoment=cos(Angle)*Moment;
YAxisMoment=sin(Angle)*Moment;
Ix=tbeam(Base,Height,BaseHeight,WebT,'n','Ix');
Iy=tbeam(Base,Height,BaseHeight,WebT,'n','Iy');
CentX=tbeam(Base,Height,BaseHeight,WebT,'n','centX');
CentY=tbeam(Base,Height,BaseHeight,WebT,'n','centY');
Xoffset=PointX-CentX;
Yoffset=CentY-PointY;
SigmaFromX=XAxisMoment*Yoffset/Ix
SigmaFromY=YAxisMoment*Xoffset/Iy
TotalStress=SigmaFromX+SigmaFromY
```

Remember the sign convention is such that tension is a positive stress. At point a:
```
TotalStress =
-1.7796e+06
```
At point b:
```
TotalStress =
1.7810e+06
```
There is a way to make this routine even better. A three dimensional representation of the stress is possible:

CH2909.m

```
Base=0.07; %meters
Height=0.1; %meters
BaseHeight=0.02; %meters
WebT=0.01; %meters
Moment=50; %Newton meters
Angle=DR(30);
XAxisMoment=cos(Angle)*Moment;
YAxisMoment=sin(Angle)*Moment;
Ix=tbeam(Base,Height,BaseHeight,WebT,'n','Ix');
Iy=tbeam(Base,Height,BaseHeight,WebT,'n','Iy');
CentX=tbeam(Base,Height,BaseHeight,WebT,'n','centX');
CentY=tbeam(Base,Height,BaseHeight,WebT,'n','centY');
x=linspace(0,Base);
y=linspace(0,Height);
[X,Y]=meshgrid(x,y);
Xoffset=X-CentX;
Yoffset=CentY-Y;
SigmaFromX=XAxisMoment*Yoffset/Ix;
SigmaFromY=YAxisMoment*Xoffset/Iy;
NormalStress=SigmaFromX+SigmaFromY;
invalid=find((X<(Base-WebT)/2 | X>(Base+WebT)/2) & (Y<Height-BaseHeight));
NormalStress(invalid)=nan;
mesh(X,Y,NormalStress)
```

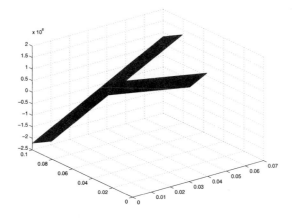

Figure 29.6   Normal stress distribution on a T-beam.

Notice on the plot where the zero stress line is, Figure 29.6. Remember stress below that line is compression. To better see the no stress plane, try plotting the absolute value of the stress. There will be a sharp fold in the diagram indicating the zero-stress line, Figure 29.7. Note that this line does go through the centroid.

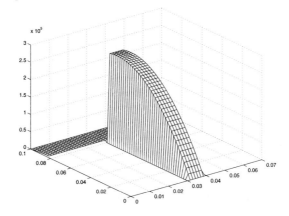

Figure 29.7   Absolute value of normal stress on T-beam.

This shows the power of MATLAB very well. The second program is scarcely longer than the first, but the output is for every point on the cross section on the beam. The key to this is that MATLAB deals with entire matrices of data values, just as well as it does with scalars that were used in the first program. The

275

routine finds the stresses on a rectangular area the size of the beam. Then the program finds all the points that lie outside the cross section and assigns a stress of NaN (Not A Number) to them so they will not appear on the plot. Similar routines could be built for all of the standard cross sectional shapes by changing the criteria of what does and does not lie on the beams cross section. The graphs look more impressive on screen with the full color.

29.6   Transverse shear

As in the last example, a three dimensional representation of the stress will be produced. Solving for a single point, Figure 29.8, has already been covered sufficiently in Chapter 26.

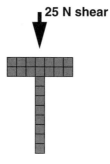

Figure 29.8   Beam cross section in shear.

CH2910.m

```
Base=0.07; %meters
Height=0.1; %meters
BaseHeight=0.02; %meters
WebT=0.01; %meters
Shear=25; %shear
Ix=tbeam(Base,Height,BaseHeight,WebT,'n','Ix');
CentY=tbeam(Base,Height,BaseHeight,WebT,'n','centY');
x=linspace(0,Base,50);
y=linspace(0,Height,50);
[X,Y]=meshgrid(x,y);
ShearStress=zeros(size(X));
BaseArea=find(Y>=Height-BaseHeight);
BaseAreaT=Base;
BaseAreaQ=abs((Y+Height)/2-CentY).*(Height-Y)*Base;
ShearStress(BaseArea)=Shear*BaseAreaQ(BaseArea)/(Ix*BaseAreaT);
WebArea=find(X>(Base-WebT)/2 &X<(Base+WebT)/2 & Y<Height-
```

```
BaseHeight);
WebAreaT=WebT;
WebAreaQ=abs((Y/2)-CentY).*Y*Base;
ShearStress(WebArea)=Shear*WebAreaQ(WebArea)/(Ix*WebAreaT);
invalid=find((X<(Base-WebT)/2 | X>(Base+WebT)/2) & (Y<Height-BaseHeight));
ShearStress(invalid)=nan;
mesh(X,Y,ShearStress)
```

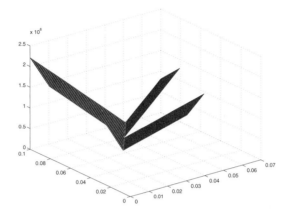

Figure 29.9   Shear stress distribution on T-beam.

        This graph, Figure 29.9, appears to have no stress in the wide area of the beam. This is not true, it is just that the stress is so large in the web that the stress in the wide area looks small in comparison. Notice how the stress grows larger as the distance from the edge grows.

29.7   Pressure loading

A typical beverage can is under pressure when purchased. By attaching a strain gauge to the can, the pressure can be estimated knowing the relationship between strain, stress, and pressure in thin-walled pressure vessels.
        The can is made of aluminum, the radius is 3 cm, the thickness is 0.001 mm and the strain gauge measurements after opening the can are:

$$\varepsilon_0 = -100\mu \qquad \varepsilon_{45} = -100\mu \qquad \varepsilon_{90} = -80\mu$$

        What is the pressure that was originally in the can if the zero angle is oriented along the length of the can?

If the strains are negated, they can be thought of as the strain caused by filling the can. These strains can be converted to a stress state. The stress state can be used to figure out the pressure by two different methods. The two pressures that are calculated can be averaged to find the true pressure in the unopened can. Remember, this is real data. It will not match the theory perfectly.

CH2911.m

```
Strains=[25 0 35]*1e-6;
Angles=DR([0 45 90]);
StrainState=rosette(Strains,Angles);
E=matprop('aluminum','E','SI')*1e9;
v=matprop('aluminum','Poissons','SI');
StressState=strain2stress(StrainState,E,v)
radius=0.03; %meters
thickness=0.001e-3;

PressureAxial=StressState(1)*2*thickness/radius
PressureHoop=StressState(2)*thickness/radius

Pressure=mean([PressureAxial PressureHoop])
```

```
Pressure =
  164.1800
```

This problem worked backwards from the normal routine of finding the stresses due to loading. However, this showed just one of the many practical applications of the routines being created here.

29.8 Combined loading

Combined loading should not be difficult if each of the loading types is considered singly then combined at the end to find the complete stress state. If the beam is 2 cm deep into the paper, see Figure 29.10, draw the diagrams for:

- Area
- Normal force and normal stress
- Shear force and shear stress at the centroid
- Bending stress on the bottom edge

Finally, find the complete stress state at the point one cm up and one cm along the beam from the lower connection to the wall, Draw the Mohr's circle for this point's stress state, see Figure 29.11:

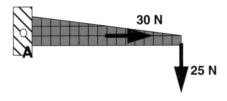

Figure 29.10   Cantilevered beam of changing cross section.

```
x=0:.001:0.16; %meters
PointAx=0.01; %meters
PointAy=0.01; %meters
Depth=0.02; %meters;
Height=diagram(x,'linear',[0.03 0.01],[0 0.16]);
Area=Height*Depth;
n(1,:)=diagram(x,'point',30,0);
n(2,:)=diagram(x,'point',-30,0.08);
Normal=sum(n);
s(1,:)=diagram(x,'point',-25,0);
s(2,:)=diagram(x,'point',25,0.16);
Shear=sum(s);
Moment=diagramintegral(x,Shear);
figure(1)
subplot(3,1,1)
plot(x,Area)
title('Area')
expandaxis; showx
subplot(3,1,2)
plot(x,Normal)
title('Normal force')
expandaxis; showx
subplot(3,1,3)
plot(x,Normal./Area)
title('Normal stress')
expandaxis, showx
Ix=Depth*Height.^3/12;
figure(2)
subplot (2,1,1)
plot(x,Shear)
title('Shear force')
expandaxis; showx
subplot(2,1,2)
plot(x,Shear.*(Height/4).*(Area/2)./(Ix*Depth))
title('Shear stress at centroid');
expandaxis; showx
figure (3)
subplot(2,1,1)
plot(x,Moment)
title('Bending Moment')
expandaxis; showx
subplot(2,1,2)
plot(x,Moment.*(Height/2)./Ix)
title('Normal stress from bending on lower edge')
expandaxis; showx
```

```
%%%
NormalStressAtA=interpolate(x,Normal./Area,PointAx)
HeightAtA=interpolate(x,Height,PointAx);
IxAtA=interpolate(x,Ix,PointAx);
ShearAtA=interpolate(x,Shear,PointAx);
QAtA=((HeightAtA/2)-(PointAy/2))*(HeightAtA*Depth);
ShearStressAtA=ShearAtA*QAtA/(IxAtA*Depth)
MomentAtA=interpolate(x,Moment,PointAx);
MomentStressAtA=MomentAtA*((HeightAtA/2)-PointAy)/IxAtA
TotalSigmaX=NormalStressAtA+MomentStressAtA;
TotalSigmaY=0;
TotalShearXY=ShearStressAtA;
StressState=[TotalSigmaX, TotalSigmaY, TotalShearXY];
figure (4)
mohrs(StressState,'plane stress')
```

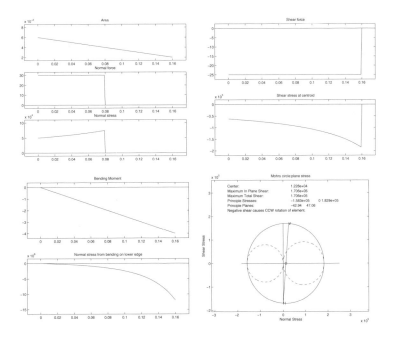

Figure 29.11   Plotted solutions to Example 29.8

30

Displacement of Beams

30.1   Introduction

Finding the displacement in a beam or shaft is a natural extension from the shear and moment diagrams that have been created in the previous chapters. In fact, it will take only one additional line of code by the end user to create the data for the displacement of the beam. Using traditional solution methods, engineers had to do lengthy calculations to find this data.

30.2   Problem

Given the moment diagram for a beam, the properties of the area moment of inertia, and the Young's modulus: find the displacement of the neutral axis at any point along the beam. An exaggerated displacement drawing is shown in Figure 30.1.

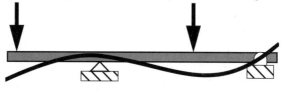

Figure 30.1   Exaggerated elastic curve.

30.3   Theory

The elastic curve, or displacement diagram, of a beam can be derived from the moment diagram of a beam. The creation of these diagrams was covered in Chapter 24.

The first integration of the moment diagram yields the slope of the beam. The second integration yields the displacement diagram. Both of these diagrams can prove useful.

$$\text{Moment} \qquad f(x)$$

$$\text{Slope} \qquad \int_0^L f(x) = g(x) + c_1$$

$$\text{Displacement} \qquad \int_0^L\int_0^L f(x) = h(x) + c_1 x + c_2$$

Eq. 30.1

These two constants of integration are free choices, they are chosen according to the boundary conditions. These boundary conditions are found from the support types used.

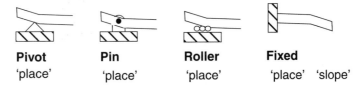

Pivot	Pin	Roller	Fixed
'place'	'place'	'place'	'place' 'slope'

Figure 30.2   Restraint types and their keywords.

The keywords in Figure 30.2 refer to inputs that must be given for the routines that have been developed. The keyword "place" is used for any support that disallows movement in the direction perpendicular to the beam. In the beam theory being used for these routines, the beam's deflection along it's own axis is considered insignificant. For this reason the roller and the pivot, which allow movement along the axis of the beam are considered equivalent to the pin, which does not. The keyword "slope" is used for any support that disallows a change in slope. The most common of these, the fixed reaction, allows neither the slope nor the position to change.

Every problem encountered in this section will be statically determinant. This means that whatever combination of supports is used, there will only be two restrictions. In practicality, this means there will either be a fixed end, or some combination of two of the others. The next chapter will cover statically indeterminate beams.

These two restrictions will dictate the constants of integration in Equation 30.1. These constants are found automatically by MATLAB and are used in the integration. The integration technique is the same trapezoid type used in the routine *diagramintegral.m*

30.4   Template

`displace.m`

```
function [D, Slope]=displace(x,Moment,type,placement,E,I)
%DISPLACE Displacement of a beam.
%    [DISPLACEMENT,SLOPE]=DISPLACE(X,MOMENT,TYPE,PLACEMENT,E,I) will find the
%    DISPLACEMENT and SLOPE of a beam that is being acted upon by the MOMENT
%    given to it.  The MOMENT data should be developed with the DIAGRAM and
%    DIAGRAMINTEGRAL routines.
%
%    TYPE describes the type of supports used, the options are:
%        ['place' 'place'] for a pin supported beam.
%        ['slope' 'place'] and ['place' 'slope'] for a beam that has a
%        restriction on it's slope at one point and its placement at that or
%        another point.  Often the place and slope restraints will be at the
%        same point for a fixed support like the wall in a cantilevered beam.
%    PLACEMENT is a two entry vector describing the place along the beam that
%        the corresponding support is acting.
%    E is the Young's modulus of the beam.
%    I is the area moment of inertia of the beams cross section.
%
%    See also DIAGRAM, DIAGRAMINTEGRAL, FIXEDFIXED, FIXEDPIN, PINPIN.

%    Details are to be found in Mastering Mechanics I, Douglas W. Hull,
%    Prentice Hall, 1999

%    Douglas W. Hull, 1999
%    Copyright (c) 1999 by Prentice Hall
%    Version 1.00

type=lower(type);

if nargin<5 E=1; end
if nargin<6 I=1; end
```

```
%type is either 'place' or 'slope', will be in a vector with two words
%placement is the distance along the x that the restraint is placed

first=type(1:5);
second=type(6:10);
if ~(strcmp(first,'place') | strcmp(first,'slope'))
  disp('Invalid restraint type, use ''slope'' or ''place''');
  return
end

if ~(strcmp(second,'place') | strcmp(second,'slope'))
  disp('Invalid restraint type, use ''slope'' or ''place''');
  return
end

if strcmp(second,'slope') & strcmp(first,'slope')
    disp('Can not deal with ''slope'' ''slope'' restraint.')
    disp('Try a redundancy routine.')
    return
end

Slope=diagramintegral(x,Moment);

if strcmp(first,'slope')
  Slope=Slope-interpolate(x,Slope,placement(1));
  D=diagramintegral(x,Slope);
  D=D-interpolate(x,D,placement(2));
elseif strcmp(second,'slope')
  Slope=Slope-interpolate(x,Slope,placement(2));
  D=diagramintegral(x,Slope);
  D=D-interpolate(x,D,placement(1));
else
  D=diagramintegral(x,Slope);
  ErrorMatrix=[-interpolate(x,D,placement(1)); -
interpolate(x,D,placement(2))];
  PlaceMatrix=[placement' [1 1]'];
  Coefs=inv(PlaceMatrix)*ErrorMatrix;
  Slope=Slope+Coefs(1);
  D=(D+Coefs(1)*x+Coefs(2));
end

Slope=Slope/(E*I);
D=D/(E*I);
```

```
function []=plotSMSD(x,shear,moment,slope,displacement)
%PLOTSMSD Plots a Shear, Moment, Slope and Displacement diagram.
%    PLOTSMD(X,SHEAR,MOMENT,SLOPE,DISPLACEMENT) is a quick routine to show
%    the SHEAR, MOMENT, SLOPE and DISPLACEMENT diagrams on the same figure.
%    This routine can and should b modified to specific needs.
%
%    See also PLOTSMD.

%    Details are to be found in Mastering Mechanics I, Douglas W. Hull,
%    Prentice Hall, 1999

%    Douglas W. Hull, 1999
%    Copyright (c) 1999 by Prentice Hall
%    Version 1.00


subplot(4,1,1)
plot ([0,x],[0,shear])
title ('Shear')
expandaxis; showx

subplot(4,1,2)
plot ([0,x],[0,moment])
title ('Moment')
expandaxis; showx

subplot (4,1,3)
plot (x,slope)
title ('Slope')
expandaxis; showx

subplot (4,1,4)
plot (x,displacement)
title ('Displacement')
expandaxis; showx
```

```
>>E=210e9;
>>I=17e-6;
>>x=0:.05:8;
>>s(1,:)=diagram(x,'point',-10,0);
>>s(2,:)=diagram(x,'point',10,8);
>>Shear=sum(s);
>>m(1,:)=diagram(x,'point',80,4);
>>m(2,:)=diagramintegral(x,Shear);
>>Moment=sum(m);
>>[Di,Sl]=displace(x,Moment,['place','place'],[0 8],E,I);
>>plotSMSD(x,Shear,Moment,Sl,Di)
```

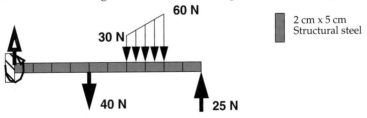

30.5   Output

Cantilevered beam: Draw the slope and displacement diagrams along with the shear and moment diagrams for the following situation, see Figure 30.3:

Figure 30.3   Displacement of a fixed beam.

The first step in this problem is to find the support reactions. For a review on this procedure, see Chapter 2.

CH3001.m

```
[f, p]=distload(-10, -30, 0.2);
af=[0 -40 .4 0; 0 25 1 0; 0 f 0.6+p 0];
[Force,Moment]=reaction(af,[0 0]);
```

The next step is to create the shear and moment diagrams, see Figure 30.4. For a review of this procedure, see Chapter 24:

```
x=0:.002:1;
s(1,:)=diagram(x,'point',Force(1,2),Force(1,3));
s(2,:)=diagram(x,'point',af(1,2),af(1,3));
s(3,:)=diagram(x,'distributed',[-10 -30],[0.6 0.8]);
s(4,:)=diagram(x,'point',af(2,2),af(2,3));
Shear=sum(s);
m(1,:)=diagram(x,'point',-Moment,0);
m(2,:)=diagramintegral(x,Shear);
Moment=sum(m);
```

Next the characteristics of the beam need to be calculated. For a review of these procedures, see Chapter 7 and Chapter 19:

```
E=matprop('structural steel','E','SI')*1e9; %Pascals
I=rectangle(0.02,0.05,'Ix');
```

Finally, the displacement and the slope need to be found and graphed:

```
[Di,Slope]=displace(x,Moment,['place','slope'],[0 0],E,I);
plotSMSD(x,Shear,Moment,Slope,Di)
```

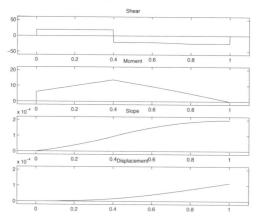

Figure 30.4   Shear, moment, slope and displacement of a beam. Overlap of title and axis label is normal in MATLAB screen shots.

Frequently when these graphs are generated, MATLAB will overlap the title of one graph with the axis of another. This can be avoided by putting less graphs in a figure, or by enlarging the window the graph is drawn in.

Simply supported beam: Show the displacement graph for the beam shown in Figure 30.5. What is the maximum deflection of the beam?

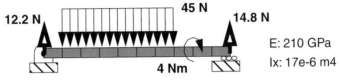

Figure 30.5   Simply supported beam.

For this problem, much of the data has already been given:

CH3002.m

```
x=0:0.01:1;
s(1,:)=diagram(x,'point',12.2,0);
s(2,:)=diagram(x,'distributed',[-45 -45],[0.1 0.7]);
s(3,:)=diagram(x,'point',14.8,1);
Shear=sum(s);
m(1,:)=diagram(x,'point',4,0.8);
m(2,:)=diagramintegral(x,Shear);
Moment=sum(m);
E=210e9; %Pascals
I=17e-6; %meters^4
Displacement=displace(x,Moment,['place','place'],[0 1],E,I);
plotSMD(x,Shear,Moment,Displacement);
disp(['The max deflection is '
num2str(max(abs(Displacement)))])
```

There are many ways to manipulate this displacement data, it is essentially the same as the methods of manipulating shear and moment that were covered in Chapter 24.

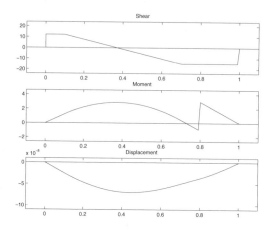

Figure 30.6   Shear, moment and displacement diagrams.

Notice in this problem, the slope data was not gathered, and consequently was not plotted. This problem gives the essence of the displacement problems, it shows how most of the work in these problems is in setting up the shear and moment diagrams. Once they are created, the displacement is easy to find. See Figure 30.6

30.6   Features

This section showed a new use for the *plotSMD.m* routine. It also gave a very simple way to find the displacement in a beam. If *I* and *E* are not known for the beam, they do not have to be entered into the routine. They act only as scaling factors. If they are not known, then the displacement curve will be good only for the shape of the deflection, not the actual values. Often times, only the displacement, not the slope, will be desired. In this case, it can simply be disregarded as in the second example of this chapter. Different plotting routines can be used if desired, *plotSMSD.m* and *plotSMD.m* are provided simply as base routines that can and should be modified to your specific needs.

30.7   Summary

Required argument, optional argument

[*displacement,* slope]=displace(*x,moment,types,place*)

Will give the displacement along a beam. *Types* is a matrix of two text strings. The text strings are either "place" meaning the beam cannot move up or down at a certain point, or "slope" meaning the beam is constrained to stay horizontal at a certain point. The place variable is a matrix of two values indicating where along the beam the two supports are. Remember, a fixed support, counts as both a "place" and a "slope" support acting at the same point.

plotSMD(*x, shear, moment, displacement*)

This routine will simplify the construction of shear, moment, and displacement diagrams. It can and should be customized to your personal needs. This represents a very generic and simple set of diagrams. Additional labeling and color changing may be desired.

plotSMSD(*x, shear, moment, slope, displacement*)

This routine will simplify the construction of shear, moment, slope and displacement diagrams. It can and should be customized to your personal needs. This represents a very generic and simple set of diagrams. Additional labeling and color changing may be desired.

interpolate(*x,y,x value*)

Given a set of x and y data that represents a function, and the desired x *value,* the routine will use linear interpolation to find the unknown y value.

31

Statically Indeterminate Beams

31.1   Introduction

Many times beams will have redundant supports. The forces provided by these supports cannot be found by using the equilibrium equations alone. Knowing the material and cross sectional properties of the beam, the remaining displacement equations can be derived.

31.2   Problem

Given a beam with any number of redundant supports, see Figure 31.1, find the forces provided by the redundant supports so that the deflection may be solved for by the *displace.m* routine.

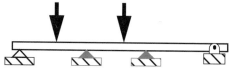

Figure 31.1   Statically indeterminate beam, redundant supports grayed.

31.3 Theory

Since the displacement of the beam is considered to be in the linear region, the method of superposition will hold. The solution technique used here is similar to the way the redundant supports were solved for in the statically indeterminate case of axial loading. The first step is to ignore all of the redundant supports and find the displacement that would occur. Since any of the supports can be considered to be redundant, the supports that will be considered redundant are the moments given by a fixed support (making them effectively a pin joint) and the pin supports that are not the outer most ones. See Figure 31.2.

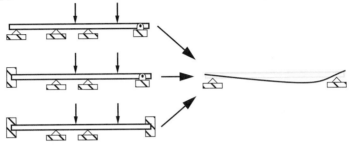

Figure 31.2   Ignoring redundant supports.

Once the redundant supports are removed, the problem can be solved for the displacement of the beam. For each support that was ignored, a force or moment must be added to make up for it. These forces and moments must be of such a magnitude that the slope is forced back to zero in the case of a fixed support, or that the displacement is forced back to zero in the case of a pin or pivot support. So in Figure 31.2 the top beam would have a force at each of the pivot supports. These two forces must be of such a magnitude that they force the displacement to be zero at the two pivots. Mathematically, for the general case of n supports it looks like this see Figure 31.3.

Figure 31.3   Notation for a beam with only pin/pivot type supports.

$$
\begin{bmatrix} Dh_1 \\ Dh_2 \\ Dh_3 \\ Dh_4 \end{bmatrix} = \begin{bmatrix} k_{11} & k_{12} & k_{13} & k_{14} \\ k_{21} & k_{22} & k_{23} & k_{24} \\ k_{31} & k_{32} & k_{33} & k_{34} \\ k_{41} & k_{42} & k_{43} & k_{44} \end{bmatrix} \begin{bmatrix} f_1 \\ f_2 \\ f_3 \\ f_4 \end{bmatrix} \qquad \text{Eq. 31.1}
$$

For a beam with one fixed end and redundant pin supports, see Figure 31.4:

Figure 31.4   Notation for a beam with a fixed end and pin supports.

$$
\begin{bmatrix} q_1 \\ Dh_1 \\ Dh_2 \\ Dh_3 \end{bmatrix} = \begin{bmatrix} k_{11} & k_{12} & k_{13} & k_{14} \\ k_{21} & k_{22} & k_{23} & k_{24} \\ k_{31} & k_{32} & k_{33} & k_{34} \\ k_{41} & k_{42} & k_{43} & k_{44} \end{bmatrix} \begin{bmatrix} m_1 \\ f_1 \\ f_2 \\ f_3 \end{bmatrix} \qquad \text{Eq. 31.2}
$$

For a beam with one fixed end and redundant pin supports, see Figure 31.5:

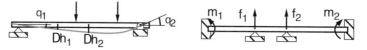

Figure 31.5   Notation for a beam with a fixed end and pin supports.

$$
\begin{bmatrix} q_1 \\ q_2 \\ Dh_1 \\ Dh_2 \end{bmatrix} = \begin{bmatrix} k_{11} & k_{12} & k_{13} & k_{14} \\ k_{21} & k_{22} & k_{23} & k_{24} \\ k_{31} & k_{32} & k_{33} & k_{34} \\ k_{41} & k_{42} & k_{43} & k_{44} \end{bmatrix} \begin{bmatrix} m_1 \\ m_2 \\ f_1 \\ f_2 \end{bmatrix} \qquad \text{Eq. 31.3}
$$

The coefficients, k, may be looked up in any standard reference book. If there are more or less redundant supports, the matrices change sizes to reflect that change.

These simultaneous equations can easily be solved for the forces and moments. Once the forces and moments are solved for, then the displacement can be solved by treating these forces and moments purely as any other external force or moment. The displacement is found by treating the problem as a beam supported by the two nonredundant pin joints nearest the ends.

The following three routines all make use of a fourth routine, *makepins.m*. All three stiffness matrices have a similar submatrix relating the deflection at the pin restraints by the forces generated at the other pin restraints. This submatrix is created with the *makepins.m* routine.

31.4   Template

`pinpin.m`

```
function [forces]=pinpin(x,s,m,a,EndSupports,E,I)
%PINPIN Redundant support forces.
%    PINPIN(X,SHEAR,MOMENT,PLACEMENT,ENDS,L,E,I) will find the redundant
%    forces supplied by any redundant pin support along the length of the
%    beam.
%
%    SHEAR is the shear acting along the beam, this should be created with
%       the DIAGRAM routine.  It does not have to be summed into a single
%       vector for use in the routine.
%    MOMENT is the moment acting along the beam, this should be only the
%       point moments created with the DIAGRAM routine.  It does not have to
%       be summed into a single vector for use in the routine.  It should not
%       include the integral of the shear as created with the routine
%       DIAGRAMINTEGRAL.
%    PLACEMENT is a vector with the location of every pin support.
%    ENDS is the placement of the two outermost pin supports along the length
%       of the beam.
%    L is the length of the beam, may extend beyond the pin supports.
%    E is the Young's modulus.
%    I is the area moment of inertia of the beam cross section.
%
%    See also DISPLACE, FIXEDFIXED, FIXEDPIN.

%    Details are to be found in Mastering Mechanics I, Douglas W. Hull,
%    Prentice Hall, 1999

%    Douglas W. Hull, 1999
%    Copyright (c) 1999 by Prentice Hall
%    Version 1.00


[ShearRows, ShearCols]=size(s);
[MomentRows, MomentCols]=size(m);

Shear=sum(s,1);
```

```
if MomentCols==1 %just sent a dummy
  Moment=diagramintegral(x,Shear);
else
  m(MomentRows+1,:)=diagramintegral(x,Shear);
  Moment=sum(m);
end;

d=displace(x,Moment,['place' 'place'],EndSupports,E,I);

for gapli=1:length(a);
  Deltas(gapli)=-interpolate(x,d,a(gapli));
end

L=EndSupports(2)-EndSupports(1);
a=a-EndSupports(1);
b=L-a;
coefs=makepins(a,L,0)/(6*E*I*L);


forces=inv(coefs)*Deltas';
```

```
>>E=210e9;
>>I=17e-6;
>>x=linspace(0,8);
>>s(1,:)=diagram(x,'point',-10,0);
>>s(2,:)=diagram(x,'point',10,8);
>>m(1,:)=diagram(x,'point',80,4);
>>redundant=pinpin(x,s,m,6,[0,8],E,I)
redundant =
        6.6620
```

80 Nm

E: 210 GPa
I_x: 17e-6 m^4

6 m

8 m

10 N   10 N

80 Nm

80 Nm

6.66 N

fixedpin.m

```
function [forces]=fixedpin(x,s,m,a,EndSupport,E,I)
%FIXEDPIN Redundant support moments and forces.
%   FIXEDPIN(X,SHEAR,MOMENT,PLACEMENT,L,E,I) will find the redundant moment
%   at the fixed support and the force supplied by any redundant pin
%   supports along the length of the beam.
%
%   SHEAR is the shear acting along the beam, this should be created with
%     the DIAGRAM routine.  It does not have to be summed into a single
%     vector for use in the routine.
%   MOMENT is the moment acting along the beam, this should be only the
%     point moments created with the DIAGRAM routine.  It does not have to
%     be suummed into a single vector for use in the routine.  It should
%     not include the integral of the shear as created with the routine
%     DIAGRAMINTEGRAL.
%   PLACEMENT is a vector with the location of every pin support.
%   L is the length of the beam, may extend beyond the last pin support.
%   E is the Young's modulus.
%   I is the area moment of inertia of the beam cross section.
%
%   See also DIAGRAM, FIXEDFIXED, PINPIN.

%   Details are to be found in Mastering Mechanics I, Douglas W. Hull,
%   Prentice Hall, 1999

%   Douglas W. Hull, 1999
%   Copyright (c) 1999 by Prentice Hall
%   Version 1.00


b=EndSupport-a;

[ShearRows, ShearCols]=size(s);
[MomentRows, MomentCols]=size(m);

if ShearRows>1
  Shear=sum(s);
else
  Shear=s;
end

if MomentCols==1 %just sent a dummy
  Moment=diagramintegral(x,Shear);
else
  m(MomentRows+1,:)=diagramintegral(x,Shear);
  Moment=sum(m);
end

[d sl]=displace(x,Moment,['place' 'place'],[0 EndSupport],E,I);

Deltas(1)=-interpolate(x,sl,0);
coefs=2*EndSupport^2/(6*E*I*EndSupport);
```

```
if a~=0
  i=1;
  for gapli=2:length(a)+1;
    Deltas(gapli)=-interpolate(x,d,a(gapli-1));
    i=i+1;
  end
  SubSca=2*EndSupport^2;
  SubCol=(a.*((a.^2)-(3*EndSupport*a)+(2*EndSupport^2)))';
  SubRow=(a.*b.*(EndSupport+b));
  SubMat=makepins(a,EndSupport,0);
  coefs=[SubSca SubRow; SubCol SubMat]/(6*E*I*EndSupport);
end

forces=inv(coefs)*Deltas';
```

```
>>E=210e9;
>>I=17e-6;
>>x=linspace(0,8);
>>s(1,:)=diagram(x,'point',-10,0);
>>s(2,:)=diagram(x,'point',10,8);
>>m(1,:)=diagram(x,'point',80,4);
>>redundant=fixedpin(x,s,m,6,8,E,I)
redundant =
    -26.6578
     17.7694
```

80 Nm

E: 210 GPa
I_x: 17e-6 m^4

6 m

8 m

10 N   10 N
      80 Nm

26.7 Nm   80 Nm

17.8 N

fixedfixed.m

```
function [forces]=fixedfixed(x,s,m,a,L,E,I)
%FIXEDFIXED Redundant support moments and forces.
%   FIXEDFIXED(X,SHEAR,MOMENT,PLACEMENT,L,E,I) will find the redundant
%   moment at the fixed supports and the force supplied by any redundant pin
%   supports along the length of the beam.
%
%   SHEAR is the shear acting along the beam, this should be created with
%      the DIAGRAM routine.  It does not have to be summed into a single
%      vector for use in the routine.
%   MOMENT is the moment acting along the beam, this should be only the
%      point moments created with the DIAGRAM routine.  It does not have to
%      be summed into a single vector for use in the routine.  It should not
%      include the integral of the shear as created with the  routine
%      DIAGRAMINTEGRAL.
%   PLACEMENT is a vector with the location of every pin support.
%   L is the length of the beam.
%   E is the Young's modulus.
%   I is the area moment of inertia of the beam cross section.
%
%   See also DISPLACE, FIXEDPIN, PINPIN.

%   Details are to be found in Mastering Mechanics I, Douglas W. Hull,
%   Prentice Hall, 1999

%   Douglas W. Hull, 1999
%   Copyright (c) 1999 by Prentice Hall
%   Version 1.00

b=L-a;

[ShearRows, ShearCols]=size(s);
[MomentRows, MomentCols]=size(m);

if ShearRows>1
  Shear=sum(s);
else
  Shear=s;
end

if MomentCols==1 %just sent a dummy
  Moment=diagramintegral(x,Shear);
else
  m(MomentRows+1,:)=diagramintegral(x,Shear);
  Moment=sum(m);
end

[d sl]=displace(x,Moment,['place' 'place'],[0 L],E,I);

Deltas(1)=-interpolate(x,sl,0);
Deltas(2)=-interpolate(x,sl,L);
coefs=[2*L^2 -L^2;-L^2 2*L^2]/(6*E*I*L);
```

```
if a~=0
  i=1;
  for gapli=3:length(a)+2;
    Deltas(gapli)=-interpolate(x,d,a(gapli-2));
    i=i+1;
  end
  SubSca=[2*L^2 -L^2;-L^2 2*L^2];
  SubCol=[(a.*((a.^2)-(3*L*a)+(2*L^2)))' -((L-a).*(((L-a).^2)-(3*L*(L-
a))+(2*L^2)))'];
  SubRow=[(a.*b.*(L+b));-(a.*b.*(L+a))];
  SubMat=makepins(a,L,0);
  coefs=[SubSca SubRow; SubCol SubMat]/(6*E*I*L);
end
forces=inv(coefs)*Deltas';
```

```
>>E=210e9;
>>I=17e-6;
>>x=linspace(0,8);
>>s(1,:)=diagram(x,'point',-10,0);
>>s(2,:)=diagram(x,'point',10,8);
>>m(1,:)=diagram(x,'point',80,4);
>>redundant=fixedfixed(x,s,m,6,8,E,I)
redundant =
      -26.6539
       -0.0503
       17.7384
```

makepins.m

```
function [matrix]=makepins(a,L,supports)
%MAKEPINS subroutine for redundancy routine.
%   MAKEPINS(A,L,SUPPORTS) creates a submatrix for the redundancy routine.
%   Not useful as a stand-alone routine.

%   Details are to be found in Mastering Mechanics I, Douglas W. Hull,
%   Prentice Hall, 1999

%   Douglas W. Hull, 1999
%   Copyright (c) 1999 by Prentice Hall
%   Version 1.00

a=a-supports(1);
b=L-a;
x=a;

for ROW=1:length(a)
  for COL=1:length(a)
    if ROW<= COL
      matrix(ROW,COL)=b(COL)*x(ROW)*(L^2-b(COL)^2-x(ROW)^2);
    else
      matrix(ROW,COL)=x(COL)*b(ROW)*(L^2-x(COL)^2-b(ROW)^2);
    end
  end
end
```

31.5   Output

Draw the shear, moment, slope and displacement diagrams for the beam when the redundant supports are ignored, and for the beam when fully constrained as shown is Figure 31.6. Express all the forces on the beam, before and after the redundant supports are considered, in a diagram created with the *showvect.m* routine.

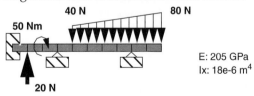

Figure 31.6   Heavily loaded statically indeterminate beam.

First, the loading must be defined, and the redundant supports identified. This will allow for the deflection to be found in the unrestrained state. See Figure 31.7.

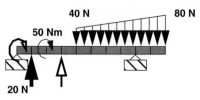

Figure 31.7   Redundant supports ignored and replaced with unknown forces.

CH3101.m

```
%%%% Setting up initial data and forces
clear
L=1; %meters
I=18e-6; %meters^4
E=205e9; %Pascals
x=linspace(0,L,500); %meters
DistMags=[-40 -80];
DistEnds=[0.4 1];
[DisFo, DisPl]=distload(DistMags(1), DistMags(2), DistEnds(2)-DistEnds(1));
af=[0 20 0.1 0;0 DisFo, 0.4+DisPl, 0];
MomentLoad=-2;
MomentPlace=0.2;
Supports=[0.3, 0.8];

%%%% Solving for the support reactions
Unknowns=[[DR(90);DR(90);0] [0; max(Supports); 0] [0; 0; 0]];
Answer=threevector(af,Unknowns,MomentLoad);

%%%% Illustrating the forces on the beam
figure(1)
showvect ([af;Answer])
title ('Forces on beam: ignoring redundant supports')
%%%% Setting up the shear and moment diagrams
%%%% Notice the s and m diagrams should not be summed prior to entry
%%%% into the fixedpin.m routine. Also the shear is not integrated
prior to
%%%% entry into the routine.
s(1,:)=diagram(x,'point',af(1,2),af(1,3));
s(2,:)=diagram(x,'point',Answer(1,2),Answer(1,3));
s(3,:)=diagram(x,'point',Answer(2,2),Answer(2,3));
s(4,:)=diagram(x,'distributed',DistMags,DistEnds);
m(1,:)=diagram(x,'point',-MomentLoad,MomentPlace);
```

```
%%%% Sending the shear and moment to the fixedpin routine. Answers broken
%%%% up into new variables.
Redundants=fixedpin(x,s,m,Supports(1),Supports(2),E,I)
RedMoment=Redundants(1);
RedForces=Redundants(2);

%%%% Drawing the SMSD diagrams as a check of the work so far. Not strictly
%%%% necisary, but a good idea.  Remember, the row of matrix m which is
%%%% the integral of shear should not be constructed before being entered
%%%% into the redundancy routines!  These routine do this automatically.
InitShear=sum(s);
m(2,:)=diagramintegral(x,InitShear);
InitMoment=sum(m);
[InDis, InSlo]=displace(x,InitMoment,['place' 'place'],[0,
max(Supports)]);
figure(2)
plotSMSD(x,InitShear, InitMoment, InSlo, InDis)
hold on
plot([0 max(Supports)],[0 0],'ko')
hold off

%%%% Finding the support reactions for the beam knowing the forces supplied
%%%% by the redundant supports
clear s m
af=[af;0 Redundants(2) Supports(1) 0];
Answer=threevector(af,Unknowns, MomentLoad+RedMoment);

%%%% Illustrating the forces on the beam
figure(3)
showvect ([af;Answer])
title ('Forces on beam: including redundant support forces')
%%%% Creating the final shear and moment diagrams
s(1,:)=diagram(x,'point',af(1,2),af(1,3)); %upward point load
s(2,:)=diagram(x,'point',af(3,2),af(3,3)); %redundant pin support
s(3,:)=diagram(x,'distributed',DistMags,DistEnds); %distributed load
s(4,:)=diagram(x,'point',Answer(1,2),Answer(1,3)); %left support
s(5,:)=diagram(x,'point',Answer(2,2),Answer(2,3)); %right support
Shear=sum(s);

m(1,:)=diagram(x,'point',-MomentLoad,MomentPlace); %Load from Moment
m(2,:)=diagram(x,'point',-RedMoment,0); % Moment from Fixed
m(3,:)=diagramintegral(x,Shear); %Integral of Shear
Moment=sum(m);

%%%% Drawing the SMSD diagrams for the final output
[Displacement, Slope]=displace(x,Moment,['place' 'slope'],[0 0]);
figure(4)
plotSMSD(x,Shear,Moment,Slope,Displacement)
hold on
plot(Supports,[0 0],'ko')
hold off
```

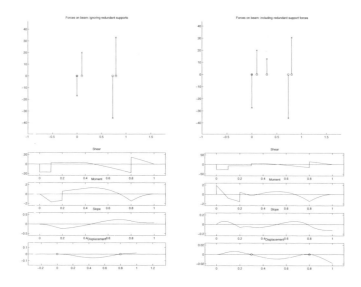

Figure 31.8   Graphed solutions to the example.

The problems appear to take a lot more code to solve, but they are no more complicated than any other type of problem, see Figure 31.8. There are just many short, simple processes to accomplish.

In summary, the steps taken in this program were:

- Set up the initial data and force vectors.
- Solve for support reactions, ignoring redundant supports.
- Illustrate forces on the beam (an optional check).
- Set up shear and moment data.
- Drawing SMSD plot, ignoring redundants (an optional check).
- Sending SM data to a routine to find redundant forces.
- Find support reactions knowing forces from redundants.
- Illustrate forces on the beam (an optional check).
- Create final shear and moment data knowing forces from redundants.
- Draw SMSD plot for the complete beam.

31.6   Features

These three routines, in combination with the many other routines that have been developed in this book allow for easy solutions to beam deflection problems, even when the beams are statically indeterminate. Even though the code gets very long in solving these problems, using paper and pencil methods would take much longer to get the same diagrams. Once the code is written, the location of the supports, and the magnitude of the forces can be changed to see very quickly what the effects are on the beam.

31.7   Summary

Required argument, optional argument

For the three redundancy routines below, *x, s, m* are all the same as they are in the context of statically determinant beams: x-axis spacing, shear, and moment. Note that neither the shears nor the moments should be summed prior to being entered into the routine. Also notice that the integral of the summed shear is not calculated prior to the routines use. *E, I* are the material and shape constants for the beam.

[forces]=pinpin(x,s,m,a,endsupports,E,I)
Returns the values of the forces supplied by redundant pin supports in a beam that is supported by only pin joints. *A* is the distance from the left end of the beam to each of the redundant supports (the redundant supports are all of those that are not the outer most pair). *Endsupports* are the distances from the left end of the beam to the outer most supports.

[forces]=fixedpin(x,s,m,a,EndSupport,E,I)
Returns the values of the forces supplied by the redundant pin supports and the moment at the fixed support. The forces vector is in the form shown at right. A is the distance from the fixed support to each of the pin supports except the last support. If there is only one pin support, enter zero for this value. *EndSupport* is the distance from the fixed support to the furthest pin support.

$$\begin{bmatrix} moment \\ force_1 \\ force_2 \\ force_n \end{bmatrix}$$

[forces]=fixedfixed(x,s,m,a,L,E,I)
Returns the values of the forces supplied by the redundant pin supports and the moments supplied at the fixed supports. The forces vector is of the form shown at the right. A is the distance from the left fixed support to each of the pin supports. If there are no pin supports, enter zero for this value.

$$\begin{bmatrix} moment_{left} \\ moment_{right} \\ force_1 \\ force_2 \\ force_n \end{bmatrix}$$

[matrix]=makepins (a,L,supports)
This is purely a function to support the three redundancy routines above. There is no direct use for it for the end user.

32

Simultaneous Equations

32.1   Introduction

Many of the problems in this book have been solved by the use of simultaneous equations. To the end user, this is for the most part invisible. However, not all situations can be covered exactly by the routines put forth in this book. In such situations the user will have to create new routines.

32.2   Examples

The use of simultaneous equations should already be familiar, the real question is how to solve them in MATLAB. For this, a few examples should be sufficient.

Structure problem: Truss problems were covered very early in this book, however, with some simple changes the problem is no longer solvable with the routines provided, see Figure 32.1.

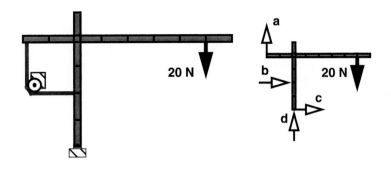

Figure 32.1   Structure not solvable by a standard routine.

In the cases studied earlier there were always three unknown forces, these could be solved for with the three equilibrium equations. In this problem, there are four unknown forces. To be solvable, there must be a fourth equation. In this situation, force a and force b must be equal for the cable to be in equilibrium. This gives the fourth equation.

$$\hat{\mathbf{A}} X = 0 = F_b + F_c$$
$$\hat{\mathbf{A}} Y = 0 = F_a + F_d - 20$$
$$\hat{\mathbf{A}} M = 0 = 2(F_a) + 2(F_b) + 5(20)$$
$$F_a = F_b$$

Eq. 32.1

These equations now have to be arranged in matrix form:

$$
\begin{bmatrix}
0 & 1 & 1 & 0 \\
1 & 0 & 0 & 1 \\
2 & 2 & 0 & 0 \\
1 & -1 & 0 & 0
\end{bmatrix}
\begin{bmatrix}
F_a \\ F_b \\ F_c \\ F_d
\end{bmatrix}
=
\begin{bmatrix}
0 \\ 20 \\ -100 \\ 0
\end{bmatrix}
$$

Eq. 32.2

To enter these into MATLAB:

```
coef=[0 1 1 0;1 0 0 1; 2 2 0 0; 1 -1 0 0];
answ=[0; 20; -100; 0];
Forces=inv(coef)*answ
Forces =
        -25
        -25
         25
         45
```

This is a very simple example showing how to use simultaneous equations in MATLAB. This takes a bit more work on the part of the user because the set-up work of developing the equations needs to be done first.

General case of parallel expansion: This is the general case of the problem from "Parallel Expansion" in Section 22.4. In this example, there will be *(N-1)* sleeves plus one bolt for a total of *N* elements. See Figure 32.2

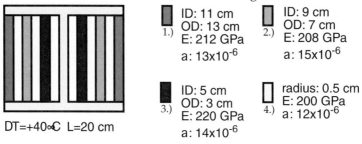

1.) ID: 11 cm OD: 13 cm E: 212 GPa a: $13x10^{-6}$

2.) ID: 9 cm OD: 7 cm E: 208 GPa a: $15x10^{-6}$

3.) ID: 5 cm OD: 3 cm E: 220 GPa a: $14x10^{-6}$

4.) radius: 0.5 cm E: 200 GPa a: $12x10^{-6}$

DT=+40∞C L=20 cm

Figure 32.2   Bolt and sleeve configuration in thermal stress.

After freely expanding, each of the sleeves will be at a certain known length *L*. When the restraining force of the bolt is applied, each of the sleeves will be pulled back through an unknown distance *d*. See Figure 32.3.

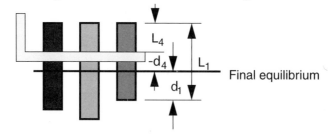

L_4 $-d_4$ L_1 d_1 Final equilibrium

Figure 32.3   Notation.

The equations of equilibrium for this case are first for the final lengths:

$$\text{Final Equilibrium Length} = L_n - d_n \qquad \text{Eq. 32.3}$$

This general equation will give $(N-1)$ equations, since the final equilibrium length is not actually known, but each equation that leads to it can be equated with every other one:

$$L_N - d_N = L_1 - d_1$$
$$L_N - d_N = L_2 - d_2 \qquad \text{Eq. 32.4}$$
$$\cdots$$
$$L_N - d_N = L_{N-1'} - d_{N-1}$$

The final equation to make the system complete is the force equation where the bolt must exert a force of compression on each of the sleeves:

$$\frac{L_1}{A_1 E_1} d_1 + \frac{L_2}{A_2 E_2} d_2 + \ldots + \frac{L_N}{A_N E_N} d_N = 0 \qquad \text{Eq. 32.5}$$

Solving each for d and putting into matrix form:

$$\begin{bmatrix} 1 & 0 & 0 & -1 \\ 0 & 1 & 0 & -1 \\ 0 & 0 & 1 & -1 \\ K_1 & K_2 & K_3 & K_4 \end{bmatrix} \begin{bmatrix} d_1 \\ d_2 \\ d_3 \\ d_4 \end{bmatrix} = L_4 - \begin{bmatrix} L_1 \\ L_2 \\ L_3 \\ L_4 \end{bmatrix} \qquad \text{Eq. 32.6}$$

$$K_n = \frac{L_n}{A_n E_n}$$

Notice the simple elegance of this solution. It can be expanded out to any number of sleeves. The pattern for expansion should be evident. To create the coefficient matrix in MATLAB, it will be easier to create three submatrices. The first is the identity matrix in the upper left-hand corner. The second is a negative ones column vector in the upper right, and the final is a stiffness row along the bottom.

Now to implement this set of equations with MATLAB:

```
E=[212 208 220 200]*1e9; %pascals
A(1)=obeam(0.13, 0.11, 'area'); %meters^2
A(2)=obeam(0.09, 0.07, 'area'); %meters^2
A(3)=obeam(0.05, 0.03, 'area'); %meters^2
A(4)=circle(0.01, 'area'); %meters^2
Alpha=[13 15 14 12]*1e-6;
DeltaT=40;
OriginalL=0.2;
%%%% End Data Entry %%% Start Matrix Cnstruction
FreeL=OriginalL*(1+(Alpha*DeltaT));

UR=eye(length(E)-1);
UL=-ones(length(E)-1,1);
K=FreeL./(A.*E);

coefs=[UR UL; K];
answ=FreeL(length(FreeL))-FreeL';

deltas=inv(coefs)*answ;

format long %Answer seen to more digits
FinalLength=FreeL'+deltas
format short %Format reset
```

This example can very easily be run again with different values, or even with more or less sleeves in the assembly.

33
Looping

33.1 Introduction

Many of the routines in this book are built for finding y as a function of x: $y = f(x)$. If a function is sufficiently complicated, it may be desirable to find trends in the relationship between the inputs and outputs of a function. Looping through several values of the input should show the trend nicely.

Another problem arises when the function is not easily reversible. For instance if a specific y is desired and the x that causes that value needs to be found. If several values for x are guessed and the values plotted, it should be easy to discover the correct value of x.

These are but a few of the scenarios where looping through different inputs could prove useful. The techniques shown here are not necessarily the most computationally efficient methods, but they are the easiest for the end user to create with little training in programming. The time saved by using a simple, but inefficient algorithm will far outweigh the saving in actual computation time for the beginning programmer.

33.2 Problem

Use the technique of looping to solve a few different types of problems:
- Trend identification
- Function inversion

33.3 Theory

A loop is a way to execute a block of code several times. There are two major types of loops in MATLAB.

Definite loops: *For* loops are definite loops because they execute the code a set amount of times. An example of this type of loop would be:

```
for n=1:10
   b=n+(n*10)
end
```

Definite loops are good for situations where you want to execute a set of instructions a specified amount of times, regardless of the results.

Indefinite loops: *While* loops are indefinite loops because they will execute a block of code as long as a certain condition is true. An example of this type of code would be:

```
b=0; n=0;
while b<80
n=n+1;
   b=n+(n*10)
end
```

The while loop is needed for times when the decision to execute the block of code again is based on the results of something that occurs in the code. For this one, the code stops once the value of *b* is too big. Notice that the value of *b* had to be defined before the first time through the loop. Notice also that the value for *n* had to be predefined before the loop, and it had to be incremented by hand each time through the loop.

For this counting scheme, the for loop looks more compact and usable. But depending on the type of problem being solved, each type of loop has it's place.

Saving values: In these examples, the value of *b* was calculated as desired, however; once the next value was calculated, the previous one was lost. In some cases this is fine, but most of the time, the data needs to be saved as the program executes. For this, an index will have to be created. This will tell MATLAB where in a vector to save the data.

CH3301.m

```
i=0; %an index
for n=0:0.1:DR(720) %counting from zero to 180 deg by a small amount
          i=i+1;
          y(i)=sin(n);
          x(i)=n;
end
plot (x,y)
```

This same data could be gathered more efficiently in other ways
```
>>x=0:0.1:DR(720);
>>y=sin(x);
>>plot (x,y)
```
However, this is meant to be an exercise in understanding loops, not in making them efficient. That is better covered by other books.

33.4   Examples

Here are a few examples that will help to illustrate the power of this method.

Finding reaction trends: This is a problem with an intuitive answer. It will be studied not because it is useful in itself, but because it can be easily understood and used for a model in more complicated problems that you may encounter.

What are the reactions at points *a* and *b* if the single force is allowed to vary in placement and magnitude as shown in Figure 33.1?

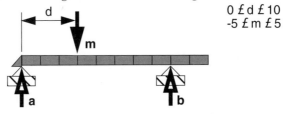

$$0 \pounds d \pounds 10$$
$$-5 \pounds m \pounds 5$$

Figure 33.1   Free body diagram.

Since there are two variables changing in this situation, the data collection will have to be in a matrix. In this matrix, each column will represent a constant value for distance, and each row will represent a constant value for magnitude. With two things changing, it is necessary that there be two loops, one inside the other. These are called nested loops. Remember that the problem will be solved like any other reaction force problem, it will just be solved many more times very quickly.

```
%%%% Bookkeeping cleaning up the workspace
clear all
close all
clc

L=10;
supports=[0 8];
unknowns=[DR(90) supports(1) 0; DR(90) supports(2) 0;0 0 0];
stepsize=0.5; %how finely the data is made
col=0; %column index
for dis=0:stepsize:10 %changing through all the distances
  col=col+1; %incrementing the column index
  row=0; %restarting the row index
  for magni=-5:stepsize:5 %nested loop for magnitude
    row=row+1; %row index
    af=[0 magni dis 0]; %defining load with new data
    reactions=threevector(af,unknowns); %finding reactions
    a(row,col)=mag(reactions(1,:),'y');
    b(row,col)=mag(reactions(2,:),'y');
    Dis(row,col)=dis;
    Magni(row,col)=magni;
  end
end
```

This quite nicely defines the reactions for over 400 scenarios. However this data is rather useless if it cannot be visualized. For that a 3-D plot can be created. The next bit of code can be used for the job of creating plots:

```
figure (1)
mesh(Dis,Magni,a)
title ('Reaction at A')
xlabel('Distance to Load')
ylabel('Magnitude of Load')

figure (2)
mesh(Dis,Magni,b)
title ('Reaction at B')
xlabel('Distance to Load')
ylabel('Magnitude of Load')
```

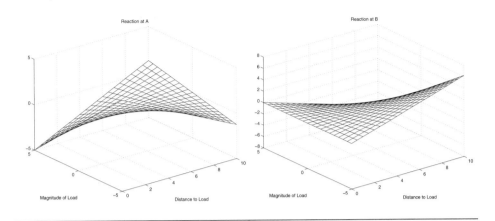

Figure 33.2 Reactions as a function of placement and magnitude of load.

These graphs, Figure 33.2, are easier to understand in full color on the computer screen. Do a quick check of these plots to see if they are valid. It is very easy to make minor errors in typing or logic and it is always a good idea to see if the answers make sense. A first easy check is to see if the value of the force at b is zero when the load is directly over a, and vice versa. Looking at the second figure, it is easy to see this is true. Upon closer inspection this is true on the first plot also.

Another check can be made by the observation that the reactions at a and b must sum to the opposite of the original load. A third figure will need to be drawn for that plot:

```
figure (3)
mesh(Dis,Magni,-(a+b))
title ('Sum of reactions')
xlabel('Distance to Load')
ylabel('Magnitude of Load')
```

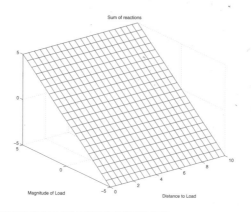

Figure 33.3   Summation of the reaction forces.

Indeed, this check also works. Notice that when the magnitude of the load is held is constant, the negative of the sum of the reactions is also constant and equal to the load, see Figure 33.3.

Data selection: It can be seen how by using this looping and graphing technique very many situations can be visualized at once. The next step is to find the subset of data that fits the selection criteria for the problem. For instance, if the support at a is to hold between 3 and 4 units of force, it would be nice to find the loading situations that fulfil this requirement. The find command is ideal for this

```
>>[valid]=find(a>=3 & a<=4);
>>[a(valid) Magni(valid) Dis(valid)]
    4.0000  -4.0000        0
    3.5000  -3.5000        0
    3.0000  -3.0000        0
    3.7500  -4.0000    0.5000
    3.2812  -3.5000    0.5000
    3.9375  -4.5000    1.0000
    3.5000  -4.0000    1.0000
    3.0625  -3.5000    1.0000
    3.6562  -4.5000    1.5000
    3.2500  -4.0000    1.5000
    3.7500  -5.0000    2.0000
    3.3750  -4.5000    2.0000
    3.0000  -4.0000    2.0000
    3.4375  -5.0000    2.5000
    3.0938  -4.5000    2.5000
    3.1250  -5.0000    3.0000
```

This gives a list of all of the loadings that fit the criteria. Note that because the routine steps through all of the loading conditions by a finite distance, there will by necessity be some valid loading conditions left out. For instance, the loading condition of 4.25 magnitude (second column) and at a distance of 1 (third column) would be a valid condition. It just happens that those specific values were never used in calculation because the step size was too coarse. For the next section, more data points will be necessary for a better graph. For this reason, the *stepsize* variable will be changed from 0.5 to 0.1 and the program run again. This is often a good technique: Write the program with a large step size so that the initial testing is fast, but then when the final data run is made, lower the step size for more precise answers.

If a graph of only the loading conditions that are valid is to be shown:

```
valid=find(a<=4 & a>=3);
A=nan*ones(size(a));
A(valid)=a(valid);
figure (4)
mesh (Dis, Magni, A)
title ('A force between 3 and 4')
xlabel('Distance to Load')
ylabel('Magnitude of Load')
```

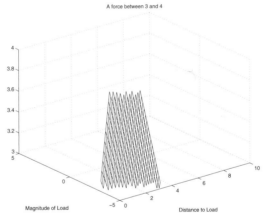

Figure 33.4   View of selected data.

This graph, Figure 33.4, does not look very useful. That is because the view is not from an intuitive place. Change to an overhead view with the command

```
>>view (2)
```

This is now a much better perspective, Figure 33.5. This view change can be done with any of the other figures. To look at the entire set of a values in the same manner:

```
>>figure (1)
>>view (2)
```

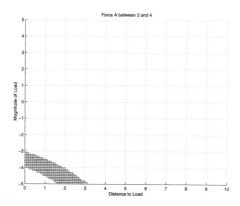

Figure 33.5   Top view of loading conditions where force *a* is between 3 and 4.

This and the other top views are much more impressive in full color on the computer monitor. Note that for this example the criteria was for a set of data that was in a certain range. Because the data is sampled at a finite spacing, it would be very unlikely to find a force value that was exactly 3.5 for example. However, looking for all the force values near 3.5 yields quite a bit of data. To zoom in on data within a smaller window near 3.5, it could possibly be necessary to decrease the step size to get more data points.

This simple example should be expandable up to more realistic problems that are encountered. For example the selection criteria for the find function could be changed to gather the data where force *a* is less than twice force *b*:

```
valid=find(a<2*b);
```

Then the rest of the graphing routines can be run with a few changes to titles. Another more complicated selection criteria might be: either of the forces is greater than four in compression or three in tension Figure 33.6:

```
valid=find(a<-4 | b<-4 | a>3 | b>3);
```

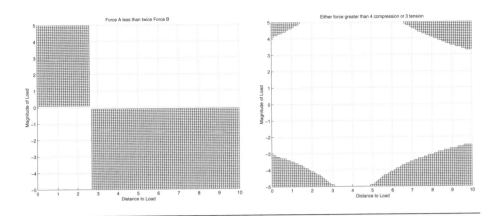

Figure 33.6   Alternative selection criteria plots.

33.5   Summary

The examples show how to construct basic loops so that many different sets of input to functions can be analyzed quickly. Once the many sets of data are gathered, it is possible to detect trends in input and output. This data can also be analyzed to find sets of input that meet certain selection criteria.

34

Redundant Beams:
Complete Examples

34.1   Introduction

The *pinpin.m* and other redundancy routines from Chapter 31 do the work of solving for the redundant forces, however as was seen in the examples for that chapter, there is a sizable amount of set-up work to be done. This chapter will create M-files to employ the redundancy routines. With these routines, the point, distributed, and moment loads simply need to be specified at the beginning of the program along with some information about the beam. The rest of the work is already laid out for any number of loads at any placement.

These three routines will all follow the same format:
- Problem specific variables are defined
- Redundant supports are ignored
- Temporary deflection is calculated
- Redundant forces and moments found from temporary deflection
- These redundant forces are applied to the beam to find the true deflection

34.2   Pin supported beams

The following is a routine for solving a beam supported by three or more pin joints. See Figure 34.1.

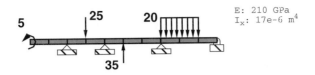

Figure 34.1   Beam deflection: pinned only.

Only the first lines of this routine need to be altered for each case. If the answer does not appear correct because the deflection is not zero through a pin connection, then increase the *NOP* (number of points).

For each of the plots, the displacement diagram will have several symbols placed along the x axis. These symbols represent the moments, load, and supports to help explain what is causing the displacements. See Figure 34.2.

CH3401.m

```
clear all %get rid of all variable and such
clc %clear the comand window
close all %close all figures

NOP=1000; %number of data points
L=10; %meters
SupportLocation=[2 4 7 10]; %meters
PointLoad=[0 -25 3 0; 0 35 5 0]; %newtons
DistPlace=[7 9]; %meters
DistMag=[-20 -20]; %newtons
MomentLoad=[5]; %newton*meters
MomentPlace=[0]; %meters
E=210e9; %Pascals
I=17e-6; %Meters^4
%%%%Don't alter below this line!%%%%

DistribLoad=dist2y(DistMag,DistPlace);
af=[PointLoad; DistribLoad];
SupportLocation=sort(SupportLocation);

if length(SupportLocation)>2
  Redundants=SupportLocation(2:length(SupportLocation)-1);
else
  error ('This routine is for redundant beams, add a third support')
end

x=linspace(0,L,NOP);

First=min(SupportLocation);
Last=max(SupportLocation);
```

```
Unknowns=[DR(90) First 0;DR(90) Last 0;0 First 0];
Reactions=threevector(af, Unknowns, sum(MomentLoad));
Left=Reactions(1,2);
Right=Reactions(2,2);


s(1,:)=diagram(x,'point',Left,First);
s(2,:)=diagram(x,'point',Right,Last);


PLShear=zeros(size(s(1,:))); %Point Loads
for gapli=1:rows(PointLoad);


PLShear(gapli,:)=diagram(x,'point',PointLoad(gapli,2),PointLoad(gapli,3));
end


DLShear=zeros(size(s(1,:))); %Distributed Loads
for gapli=1:rows(DistMag)


DLShear(gapli,:)=diagram(x,'distributed',DistMag(gapli,:),DistPlace(gapli,:));
end


TS=[s;PLShear;DLShear]; %Total Shear


for gapli=1:length(MomentLoad)
  m(gapli,:)=diagram(x,'point',-
MomentLoad(gapli),MomentPlace(gapli));
end


RedundantForces=pinpin(x,TS,m,Redundants,[First Last],E,I);


MLoad=summoment(af,[First,0]); %Moment from Load
MMoment=sum(MomentLoad); %Moment from Moment Load
MRFLoad=sum(RedundantForces.*(Redundants'-First)); %Moment from Redundant Force


right=-(MLoad+MMoment+MRFLoad)/(Last-First);
left=-mag(sumforce(af),'y')-right-sum(RedundantForces);


for gapli=1:length(RedundantForces);


RFShear(gapli,:)=diagram(x,'point',RedundantForces(gapli),Redundants(gapli));
end


RFShear=sum(RFShear,1); %Redundant Force Shear
PLShear=sum(PLShear,1); %Point Load Shear
DLShear=sum(DLShear,1); %Distributed Load Shear


clear s
s(1,:)=diagram(x,'point',left,First);
```

```
s(2,:)=diagram(x,'point',right,Last);

Shear=RFShear+PLShear+sum(s)+DLShear;
clear m
m(1,:)=diagramintegral(x,Shear);
for gapli=3:length(MomentLoad)+2
  m(gapli,:)=diagram(x,'point',-MomentLoad(gapli-2),MomentPlace(gapli-2));
end
Moment=sum(m);

[d sl]=displace(x,Moment,['place' 'place'],[First Last],E,I);

figure(1)
clf
plotSMSD(x,Shear,Moment,sl,d)
hold on

plot (SupportLocation,zeros(size(SupportLocation)),'ko')
plot
(Redundants,zeros(size(Redundants)),'ro',PointLoad(:,3),0,'g*')
plot (DistPlace,zeros(size(DistPlace)),'y*')
plot (MomentPlace,zeros(size(MomentPlace)),'mp')
hold off
```

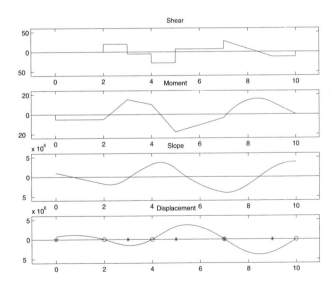

Figure 34.2   Answer to the pinned example.

34.3   Fixed and pin supports

This routine is for beams that are fixed at one end and have any number of pin supports along their lengths as shown in Figure 34.3. Figure 34.4 provides diagrams.

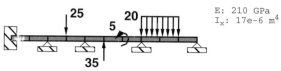

Figure 34.3   Beam fixed at one end with pins along its length.

CH3402.m

```
clear all %get rid of all variable and such
clc %clear the comand window
close all %close all figures

NOP=2000; %number of data points
L=10; %meters
SupportLocation=[2 4 7 10]; %meters
PointLoad=[0 -25 3 0; 0 35 5 0]; %newtons
DistPlace=[7 9]; %meters
DistMag=[-20 -20]; %newtons
MomentLoad=[5]; %newton*meters
MomentPlace=[6]; %meters
E=210e9; %Pascals
I=17e-6; %Meters^4
%%%%Don't alter below this line!%%%%

DistribLoad=dist2y(DistMag,DistPlace);
af=[PointLoad; DistribLoad];
SupportLocation=sort(SupportLocation);

if length(SupportLocation)>1  %The last support is not redundant
   Redundants=SupportLocation(1:length(SupportLocation)-1);
else
   Redundants=0;
end

x=linspace(0,L,NOP);

Last=max(SupportLocation);
```

```
Unknowns=[DR(90) 0 0;DR(90) Last 0;0 0 0];
Reactions=threevector(af, Unknowns, sum(MomentLoad));
Left=Reactions(1,2);
Right=Reactions(2,2);

s(1,:)=diagram(x,'point',Left,0);
s(2,:)=diagram(x,'point',Right,Last);

PLShear=zeros(size(s)); %Point Loads
for gapli=1:rows(PointLoad);

PLShear(gapli,:)=diagram(x,'point',PointLoad(gapli,2),PointLoad(gapli,3));
end

DLShear=zeros(size(s)); %Distributed Loads
for gapli=1:rows(DistMag)

DLShear(gapli,:)=diagram(x,'distributed',DistMag(gapli,:),DistPlace(gapli,:));
end

TS=[s;PLShear;DLShear]; %Total Shear

for gapli=1:length(MomentLoad)
  m(gapli,:)=diagram(x,'point',-
MomentLoad(gapli),MomentPlace(gapli));
end

answer=fixedpin(x,TS,m,Redundants,Last,E,I);

RedundantMoment=answer(1);

RedundantForces=0;
if length(answer)>1
  RedundantForces=answer(2:length(answer));
end

MLoad=summoment(af); %Moment from Load
MMoment=sum(MomentLoad); %Moment from Moment Load
MRFLoad=sum(RedundantForces.*Redundants'); %Moment from Redundant Forces
MRMoment=answer(1); %Moments from Redundant Moments

right=-(MLoad+MMoment+MRFLoad+MRMoment)/Last;
left=-mag(sumforce(af),'y')-right-sum(RedundantForces);

for gapli=1:length(RedundantForces);

RFShear(gapli,:)=diagram(x,'point',RedundantForces(gapli),Redundants(gapli));
end
```

325

```
RFShear=sum(RFShear,1); %Redundant Force Shear
PLShear=sum(PLShear,1); %Point Load Shear
DLShear=sum(DLShear,1); %Distributed Load Shear

clear s
s(1,:)=diagram(x,'point',left,0);
s(2,:)=diagram(x,'point',right,Last);

Shear=RFShear+PLShear+sum(s)+DLShear;
clear m
m(1,:)=diagram(x,'point',-RedundantMoment,0);
m(2,:)=diagramintegral(x,Shear);
for gapli=3:length(MomentLoad)+2
  m(gapli,:)=diagram(x,'point',-MomentLoad(gapli-
2),MomentPlace(gapli-2));
end
Moment=sum(m);

[d sl]=displace(x,Moment,['place' 'slope'],[0 0],E,I);

figure(1)
clf
plotSMSD(x,Shear,Moment,sl,d)
hold on
[blah, yada, LoadX, foo]=breakup(af);
plot
(0,0,'rd',Redundants,zeros(size(Redundants)),'ro',LoadX,0,'g*')
plot (SupportLocation,zeros(size(SupportLocation)),'ko')
plot (DistPlace,zeros(size(DistPlace)),'y*')
plot (MomentPlace,zeros(size(MomentPlace)),'mp')
hold off
```

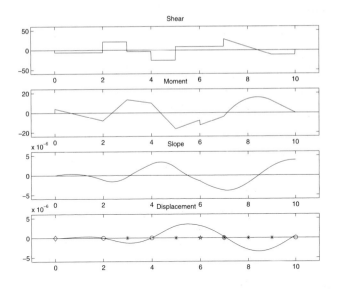

Figure 34.4   Answer to fixed pin example.

34.4   Fixed ends with pin supports

This routine is for beams that are fixed at both ends and have any number of pin supports along their lengths as in Figure 34.5. Figure 34.6 provides diagrams.

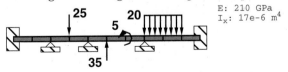

Figure 34.5   Beam fixed at both ends with pins along its length.

```
clear all %get rid of all variable and such
clc %clear the comand window
close all %close all figures

NOP=3000; %number of data points
L=10; %meters
SupportLocation=[2 4 7]; %meters
PointLoad=[0 -25 3 0; 0 35 5 0]; %newtons
DistPlace=[7 9]; %meters
DistMag=[-20 -20]; %newtons
MomentLoad=[5]; %newton*meters
MomentPlace=[6]; %meters
E=210e9; %Pascals
I=17e-6; %Meters^4
%%%%Don't alter below this line!%%%%

DistribLoad=dist2y(DistMag,DistPlace);
af=[PointLoad; DistribLoad];
Redundants=sort(SupportLocation);

x=linspace(0,L,NOP);

Unknowns=[DR(90) 0 0;DR(90) L 0;0 0 0];
Reactions=threevector(af, Unknowns, sum(MomentLoad));
Left=Reactions(1,2);
Right=Reactions(2,2);

s(1,:)=diagram(x,'point',Left,0); %Support Reactions
s(2,:)=diagram(x,'point',Right,L);

PLShear=zeros(size(s)); %Point Loads
for gapli=1:rows(PointLoad);

PLShear(gapli,:)=diagram(x,'point',PointLoad(gapli,2),PointLoad(gapli,3));
end

DLShear=zeros(size(s)); %Distributed Loads
for gapli=1:rows(DistMag)

DLShear(gapli,:)=diagram(x,'distributed',DistMag(gapli,:),DistPlace(gapli,:));
end

TS=[s;PLShear;DLShear]; %Total Shear

for gapli=1:length(MomentLoad)
  m(gapli,:)=diagram(x,'point',-MomentLoad(gapli),MomentPlace(gapli));
end

answer=fixedfixed(x,TS,m,Redundants,L,E,I);

RedundantMomentLeft=answer(1);
RedundantMomentRight=answer(2);
```

```
RedundantForces=0;
if length(answer)>2
  RedundantForces=answer(3:length(answer));
end

MLoad=summoment(af); %Moment from Load
MMoment=sum(MomentLoad); %Moment from Moment Load
MRFLoad=sum(RedundantForces.*Redundants'); %Moment from Redundant Forces
MRMoment=sum(answer(1:2)); %Moments from Redundant Moments

right=-(MLoad+MMoment+MRFLoad+MRMoment)/L;
left=-mag(sumforce(af),'y')-right-sum(RedundantForces);

RFShear=zeros(size(s(1,:))); %Redundant Force Shear
for gapli=1:length(RedundantForces);

RFShear(gapli,:)=diagram(x,'point',RedundantForces(gapli),Redundants(gapli));
end

RFShear=sum(RFShear,1); %Redundant Force Shear
PLShear=sum(PLShear,1); %Point Load Shear
DLShear=sum(DLShear,1); %Distributed Load Shear

clear s
s(1,:)=diagram(x,'point',left,0);
s(2,:)=diagram(x,'point',right,L);

Shear=RFShear+PLShear+sum(s)+DLShear;
clear m
m(1,:)=diagram(x,'point',-RedundantMomentLeft,0);
m(2,:)=diagram(x,'point',-RedundantMomentRight,L);
m(3,:)=diagramintegral(x,Shear);
for gapli=4:length(MomentLoad)+3
  m(gapli,:)=diagram(x,'point',-MomentLoad(gapli-3),MomentPlace(gapli-3));
end
Moment=sum(m);

[d sl]=displace(x,Moment,['place' 'slope'],[0 0],E,I);

figure(1)
clf
plotSMSD(x,Shear,Moment,sl,d)
hold on
plot (0,0,'rd',L,0,'rd',af(:,3),0,'g*')
if Redundants~=0
  plot (Redundants,zeros(size(Redundants)),'ko')
end
plot (DistPlace,zeros(size(DistPlace)),'y*')
plot (MomentPlace,zeros(size(MomentPlace)),'mp')
hold off
```

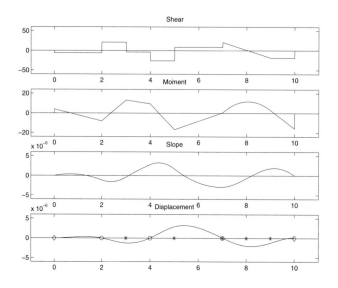

Figure 34.6   Answer to fixed ends example.

Index